MEASURING AND VISUALIZING SPACE IN ELEMENTARY MATHEMATICS LEARNING

Measuring and Visualizing Space in Elementary Mathematics Learning explores the development of elementary students' understanding of the mathematics of measure and demonstrates how measurement can serve as an anchor for supporting a deeper understanding of number operations and rational numbers.

The concept of measurement is centrally implicated in a number of mathematical operations, yet it is not often given the placement it deserves in the elementary mathematics curriculum. By drawing on K-5 classroom research, authors Lehrer and Schauble have been able to articulate a learning progression that describes benchmarks of student learning about measure in length, angle, area, volume, and rational number, exploring related concepts, classroom experiences, and instructional practices at each stage.

Offering a unique, research-driven resource for helping students develop a deep understanding of measurement to enhance mathematical understanding, as well as further learning in other STEM disciplines, the book will be relevant for scholars, teacher educators, and specialists in math education.

The book is accompanied by online resources developed for practitioners, including instructional guides, examples of student thinking, and other teacher-focused materials, helping to clarify how to bring concepts of measure and rational number to life in classrooms.

Richard Lehrer is a Professor of Education in the Department of Teaching and Learning at Vanderbilt University, Nashville, Tennessee, USA.

Leona Schauble is a Professor of Education in the Department of Teaching and Learning at Vanderbilt University, Nashville, Tennessee, USA.

MEASURING AND VISUALIZING SPACE IN ELEMENTARY MATHEMATICS LEARNING

Richard Lehrer
Leona Schauble

Routledge
Taylor & Francis Group

NEW YORK AND LONDON

Designed cover image: © Trisha Flores

First published 2023
by Routledge
605 Third Avenue, New York, NY 10158

and by Routledge
4 Park Square, Milton Park, Abingdon, Oxon, OX14 4RN

Routledge is an imprint of the Taylor & Francis Group, an informa business

© 2023 Richard Lehrer and Leona Schauble

The right of Richard Lehrer and Leona Schauble to be identified as
authors of this work has been asserted in accordance with sections
77 and 78 of the Copyright, Designs and Patents Act 1988.

Library of Congress Cataloging-in-Publication Data
Names: Lehrer, Richard, author. | Schauble, Leona, author.
Title: Measuring and visualizing space in elementary mathematics learning /
Richard Lehrer, Leona Schauble.
Description: New York, NY : Routledge, 2023. | Includes bibliographical
references and index.
Identifiers: LCCN 2022039860 (print) | LCCN 2022039861 (ebook) | ISBN
9781032262734 (hardback) | ISBN 9781032262727 (paperback) | ISBN
9781003287476 (ebook)
Subjects: LCSH: Geometry--Study and teaching (Elementary)
Classification: LCC QA461 .L466 2023 (print) | LCC QA461 (ebook) | DDC
372.7/6--dc23/eng20230110
LC record available at https://lccn.loc.gov/2022039860
LC ebook record available at https://lccn.loc.gov/2022039861

ISBN: 9781032262734 (hbk)
ISBN: 9781032262727 (pbk)
ISBN: 9781003287476 (ebk)

DOI: 10.4324/9781003287476

Typeset in Bembo Std
by KnowledgeWorks Global Ltd.

Access the Support Material: www.routledge.com/9781032262727

CONTENTS

About the Authors   *x*

Acknowledgments   *xi*

1 Measure is Fundamental   1
   Developing a Theory of Measure 2
   Relation to Existing Scholarship 8
   Organization of the Book 12
   References 13

2 The Context, Goals, and Design of the Research   17
   Research Sites and Participants 17
   Design Research: Engineering Learning to Support Its Study 18
   The Educational Design 19
      Professional Development Collaboration with Participating
         Teachers 20
      Learning Constructs 22
      Supporting Curriculum Units 24
      Assessment System 26
      Digital Tools for Collecting, Displaying, and Interpreting
         Student Data 29
   Studies of Student Learning 33
      Exploratory Learning Studies 33
      Yearly Interview Data 34
      Formative Assessment Tasks and In Situ Observations 36
   Studies of Teacher Learning and Practice 36
   References 38

3 Origins of Quantitative Reasoning in the Measure
   of Length   41
   Overview of Children's Understanding of Length 43
   Benchmarks in Thinking About Length 44
      Directly Comparing Magnitudes 44
      Explaining How Properties of Units Affect Measure 48
      Unit Iteration and Constructing a Measurement Scale 52
      2-Splitting and Symbolizing 2-Split Units as Measures 55
      3-Splitting and Symbolizing 3-Split Units 59
      Generalizing Relationships Among Units and Measures 61
   References 64

4 Creating New Quantities in the Dynamic Generation
   of Area   67
   Two Coordinated Perspectives on Measure 67
   Integrating the Two Perspectives on Measure 68
      Direct Comparison of Magnitudes 68
      *Comparing Magnitudes of Area Indirectly Through Dissection
         and Unit Dissection 68*
      Properties of Units of Area Measure 72
      Dynamic Generation of Area and Product 76
      Guided Reinvention of Area Measure Formulas 83
   References 86

5 Extending Motion to Three Dimensions: Volume
   and Its Measure   88
   Structuring and Dynamic Approaches to Volume Measure 88
      Volume Conceived as Space Inside 90
      Measuring Volume by Accumulating Units 91
      Visualizing Volume as Composites of Layers 93
      Finding Volumes of Prisms with Fractional Dimension 96
      Generating Volume Dynamically 97
   References 101

6 Integrating Figure and Motion in the Measure of Angle   104
   Dynamic and Figural Perspectives 104
   Benchmarks in Thinking About Angle 108
      Noticing Canonical Examples 108
      Representing Angles-as-Figures, Angles-as-Turns 108
      *Integrating Angle-as-Turn with Angle-as-Figure, Interior vs.
         Turn Angles 112*

Generating and Justifying Angle Measure Theorems 116
Developing New Understandings of Figures and Structures via
  Angle Theorems 120
References 122

7 Measurement Models of Arithmetic Operations
  and Rational Number   124
Initial Resources for Reasoning About Rational Number in
  Measure 125
  Unit Iteration 125
  Length Measure as a Point Along a Path 126
  Measure-Magnitude Distinction 126
  Symbolizing Measure 126
  Additive and Multiplicative Comparison 127
Rational Numbers as Measured Quantities 127
  Two-Split of a Unit Length and Half-Unit Iteration 127
  Four- and Eight-Splits of a Unit Length and Measures in
    Fourth-Unit and Eighth-Unit 129
  Three-Splits and Compositions of Two- and Three-Splits of a
    Unit Length 131
Fractions as Operators on Measured Quantities 132
  Initial Steps in Developing a Sense of Fraction as Operator 133
  Extending the Reach of Fraction-as-Operator to Refine Measure 134
  Extending Multiplication and Multiplicative Comparisons 135
References 138

8 Highlights of Student Learning Research   140
The Two Phases of Research 140
Phase I: Student Conceptions of Measure as Indicated by Yearly
  Interviews 142
  Early-Developing Conceptions of the Measure of Length 143
    Direct comparison 144
    Tiling, unit, and iteration 145
    Unit iteration and symbolizations of unit on scale 147
    Equipartitioning fractured units 148
    Measurement arithmetic 148
  Conceptions of Area Measure 149
    Necessary conditions for area 150
    Unit structuring of area 151
    Differentiating area and length measure conceptually and
      symbolically 152

Conceptions of Angle Measure 154
  Embodied turns in walking paths in primary grades 154
  Conceptions of angle and measure in later grades 155
  Understandings of angle theorems 156
Conceptions of Volume Measure 158
  *Strategies employed to measure prisms constructed of cubic
    units 158*
  *Strategies employed to measure prisms with partial
    structuring 159*
  Volume of pentagonal prism 159
  Cavalieri's principle 160
*Phase II: Summative, Formative, and In-Situ Evidence of Student
  Learning 161*
Summative Assessment 161
*Formative Assessment and In-Situ Evidence of Student
  Learning 164*
  *Guiding instruction by monitoring conceptual
    development 164*
  Formative assessment and dialogic space 167
Reflections and Prospects 173
References 173

9 Teacher Learning   175
Initiating and Sustaining a Teacher-Researcher Partnership 176
  Supporting Development of a Shared Professional Vision 177
    Constructs 177
    Curriculum 179
    Digital observation tools 180
  *Activity Structures That Forge a Professional Learning
    Community 182*
    Learning labs 183
    Mathematical investigations 187
    Communal critique 188
Investigations of Change in Professional Practice 188
  Teacher Noticings During Video Episodes 189
    Video episodes 189
    Interview procedure 191
    Interview analysis 191
    Interview results 192
  Teacher Perspectives on Professional Development 197
    Teachers' views of learning labs 198

Teachers' views of mathematical investigations 199
Teachers' views of communal critique 200
Changes in Teachers' Construct-Centered Judgments About
Students' Ways of Thinking 201
Professional Vision as a Fulcrum for Learning Progression 203
References 205

10 Measures and Models in Elementary Science   207
Characterizing Growth 210
Describing Change by Determining Differences in Quantity 210
Describing Change with Intensive Quantity: Differences in
Rates 212
Describing Change in Population with Measures of
Distribution 216
Cultivating Distributional Thinking 220
Initial Steps 220
Revisiting Cause and Chance in Generating Variability 225
Melding Chance and Cause: Investigating Precision of
Measure 228
Modeling Measurements as Signal and Noise 232
Expanding the reach of modeling chance 235
Modeling Broadens the Scope of Measures 236
References 237

Index   *240*

ABOUT THE AUTHORS

Richard Lehrer and **Leona Schauble** have been professors at the University of Wisconsin–Madison and Vanderbilt University. In long-term collaborations with public elementary and middle school educators, they seek to both understand and support the development of students' mathematical and scientific learning. A premise of this research is that understanding the development of thinking requires understanding the conditions, especially instruction, under which it is supported. Conversely, efforts to support learning are enhanced when they are guided by knowledge about how thinking develops. Thus, studies of student learning can inform and contribute to teachers' shared goals and means for improving instruction. In turn, teachers' instructional repertoire and professional vision establish the conditions for supporting new forms of student reasoning.

ACKNOWLEDGMENTS

The research and development that this book describes were supported by the National Science Foundation under grants numbered 1621088, 1316312, and 0638253. Any opinions, findings, and conclusions or recommendations expressed in this material are those of the authors and do not necessarily reflect the views of the National Science Foundation.

The graphics were rendered by Lee Druce, who worked the dross of our sketches to develop polished images of ideas at play.

We thank the children, teachers, and school administrators who have been steadfast partners with us on the extended journey that produced this volume. They have been a continual source of inspiration, dedication, and ingenuity. We appreciate the many opportunities to learn with and from our colleagues, especially David W Henderson and Daina Taimina, who became essential guides for all of us to the mathematics of space; Corey Brady for his design of software that made ongoing formative assessment of student learning feasible and fruitful; Mark Wilson and Perman Gochyyev for innovations in psychometrics that bridged the gap between ongoing assessment of learning in classrooms and more traditional summative tests; David Clarke, who generously shared his perspectives on lexicons of learning and teaching that crossed national boundaries; Patrick Thompson for his continued and generous intellectual contributions to our shared interest in the development of quantitative reasoning; and, finally, our most recent collaborators in classroom studies of the development of children's mathematical and scientific reasoning: Portia Botchway, Amanda Dickes, Linda Jaslow, Ryan "Seth" Jones, Min-Joung Kim, Marta Kobiela, Eve Manz, Joanne Mulligan, Vaughn Prain, Russell Tytler, Peta White, Panchompoo Wisittanawat, and Megan Wongkamalasai.

1

MEASURE IS FUNDAMENTAL

Measurement is fundamental to the operation of commerce, the pursuit of craft, and the expansion of knowledge in STEM professions. Yet, the very ubiquity of measurement, along with its entrenchment in commonplace tools, conceals the reality that every measure is the result of an ongoing dialogue between imaginative reach and practical grasp. Measuring involves imagining and defining measurable qualities of the world, as well as developing or using tools based on these qualities to construct a measure.

Consider, for a moment, a measure that was articulated thousands of years ago by the Greeks—namely, the length of the earth's circumference. One might imagine the earth as a sphere and, accordingly, visualize paths on it as great circles: paths that cut the sphere into two congruent parts. But having formulated this image, how can the distance of such a path be brought into practical grasp?

Eratosthenes of Cyrene knew that at the summer solstice, the longest day of the year, the sun illuminated the bottom of a well in Syene (near Aswan, Egypt) without casting any shadow. The implication was that the sun's rays must be perpendicular to the bottom of the well. (If you use a pencil and a flashlight to experiment with the position of the light and note the shadow cast by the pencil, you, like Eratosthenes, may observe that when the source of light is directly overhead, there is no shadow.) Having made this observation in Syene, Eratosthenes measured the angle of a shadow cast by a vertical pole at the same time (noon) and on the same day in Alexandria, a location nearly due north of Syene. Imagining the sun's rays as parallel lines, he could relate the angle of the shadow cast by a vertical stick in Alexandria to the portion of the circumference of a great circle accounted for by the distance between Syene and Alexandria, as illustrated in Figure 1.1.

DOI: 10.4324/9781003287476-1

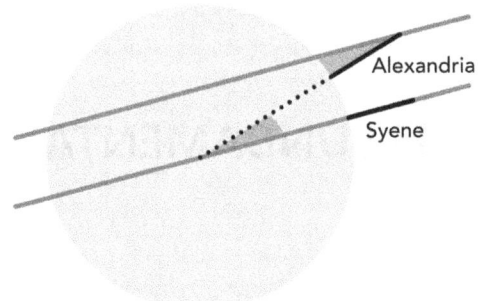

FIGURE 1.1 Relating angles to the circumference of the earth

The angle of the shadow cast in Alexandria was approximately 7.2 degrees, $\frac{1}{50}$ th of the arc of a circle. Alexandria was 5,000 stadia, each the length of a sports stadium, away from the well at Syene; this measure had been generated by members of a surveyor guild who had learned to regulate their paces to produce uniform strides. Eratosthenes concluded that the measure of the circumference of the earth must be 50 times as long as the measure of the distance between Syene and Alexandria. By most accounts (American Physical Society, 2006), his estimate of the earth's circumference was within 2% of contemporary estimates (about 25,000 miles, or 40,000 km).

Eratosthenes' measure of an attribute of the earth illustrates the inherent connection between, on the one hand, imaginatively re-representing phenomena to capture aspects considered theoretically important and, on the other hand, solving the empirical and practical problems entailed in transforming these imagined qualities into one or more measures. Erathosthenes transformed a world of sunshine, marching guild members, and shadow sticks into a model that afforded the coordination of practical measures of length and angle to determine a distance that could only be imagined.

Developing a Theory of Measure

The work that we describe in this volume proceeds from the assumption that children, too, should experience measure first-hand as an interchange between imagination and pragmatic activity. Thompson (2011) describes this interchange as a dialectic between students' conceptualizing a quantity (e.g., amount of twist) and conceptualizing a measure of it (e.g., torque, a product of force and distance). Unfortunately, most instruction in the U.S. emphasizes the pragmatic aspect of measure by focusing on how to read measurement tools, often shortchanging or entirely overlooking the role of imagination (Smith, Males, & Gonulates, 2016; Smith, Males, Dietiker, Lee, & Mosier, 2013; Thompson & Preston, 2004). Simply learning how to read tools like rulers and protractors is insufficient for developing the network of interrelated concepts that we refer to as a theory of measure (Lehrer, 2003). The constituent concepts include experience with and understanding of properties of units of measure (Crosby, 1997; Davydov & Tsvetkovich, 1991; Piaget, Inhelder, & Szeminksa, 1960) and, in addition, the relations among these properties. For example, Figure 1.2 illustrates the beginning of a conceptual system for the measure of length—the kind of knowledge a child might develop with instructional support in the primary grades. The nodes in the figure refer to central concepts, such as unit iteration (translating a unit) and measurement scale; supporting conceptual and procedural elements are depicted by links. In this case, children are envisioned as understanding the scope of iteration, ranging from units to units-of-units and to fractured units, complemented by the symbolization of iteration as a scale of measure and by a metaphor of one-dimensional motion to describe locations on this scale as a path.

Figure 1.2 represents but one part of a comprehensive theory of measure, a theory assembled by coordinating knowledge "pieces," each of which enables performance in particular contexts of measure, to expand the scope

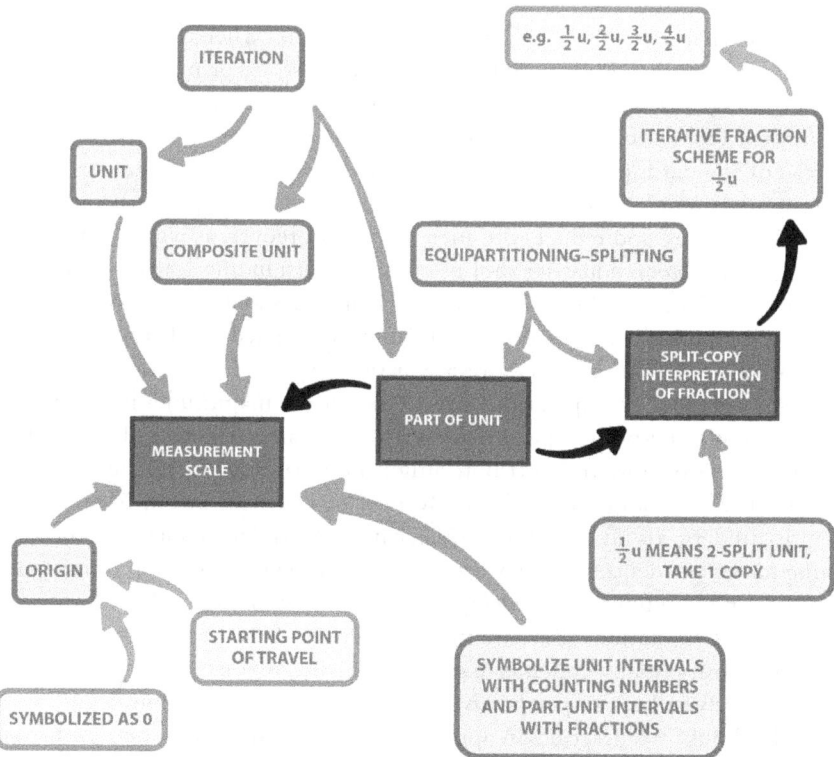

FIGURE 1.2   A network showing relations between iteration and measurement scale

of performance across multiple contexts and situations of measure (Barth-Cohen & Wittmann, 2016; diSessa, 1988; diSessa & Sherin, 1998; Wagner, 2006). Because measure is so complex and extends across so many contexts, developing a theory of measure is a goal that is achieved gradually over the elementary school years—and far beyond, once we consider the centrality of measure in disciplines from economics to physics (e.g., Crosby, 1997; Smith, Wiser, Anderson, & Krajcik, 2006; Thompson & Preston, 2004; van Fraassen, 2012). Supporting students' initial and subsequent development of this theory requires unpacking and addressing many contributing concepts, as well as supporting students as they explore how these ideas cohere into an integrated conceptual system that supports pragmatic ends.

Measurement tools like rulers and weight scales are constructed for efficiency and standardization of use, but as we noted earlier, their very efficiencies tend to efface the mathematical and phenomenological motivations that originally guided their design. Unless instruction focuses explicitly on those motivations and their underlying justifications, children may easily misunderstand the assumptions that undergird the design of the tool. The result is that students often end up misconceiving measurement merely as a list of rules or procedures for using tools (Lehrer, Jenkins, & Osana, 1998; Smith et al., 2013; Stephan & Clements, 2003). Of course, there are times when the pragmatics of use are what matters. For example, at the grocer, we are primarily concerned that the weight of the grocery scale is accurate and may not think about the conceptual foundations of its construction or the politics of its use (a state's Bureau of Standards presumably protects us from our naiveté). But instruction that singularly concentrates on efficiencies of use tends to rupture the dialogic nature of measure and reduce it to mere ritual. Moreover, as we will argue in subsequent chapters, equating tool use with understanding leads to many of the errors evident in children's responses to common standardized assessments.

Measure can also enhance and contribute to related mathematical ideas and systems. For example, measure is a critical origin of quantitative reasoning. The very idea of a quantity's size or extent (its magnitude) is grounded in the interplay between a person's conception of an attribute and his or her understanding of how that attribute might be measured (Cartwright, 1999; Crosby, 1997; Thompson, Carlson, Byerley, & Hatfield, 2014). Measure of quantity undergirds invention in engineering and inquiry in science, although young science students too rarely are encouraged to either develop or critique measures and so often fail to develop a strong understanding of their purposes, qualities, and limitations. Because measurement bridges theory and empirical practice across so many contexts (e.g., Lehrer, Kim, & Jones, 2011), we can ill afford a purely utilitarian perspective on measurement instruction.

In this book, we describe an approach to measurement instruction that is oriented toward long-term conceptual development of related concepts, tools, and methods of measure, and that delineates connections between measure and other central topics in early mathematics, such as arithmetic operations and rational number. Our focus is on measures of space, in part because length, area, volume, and angle can be literally grasped as well as imagined, and in part because once they are understood, measurement models can then enliven other mathematical systems (e.g., what is the meaning of multiplication?). We outline an instructional approach in which conceptions and practices of measure are cultivated systematically across the span of elementary education. In this instructional system, concepts are built consistently and systematically on two related grounding metaphors that leverage children's familiar, everyday intuitions and experiences about space (Lakoff & Núñez, 2000). We regard these metaphors and their blends as starting points for anchoring the intelligibility of measure and for providing practical means for extending the imaginative reach of measure.

One of these foundational metaphors involves using a rigid object, such as a thumb or footstep, to span a distance in order to get some sense of the amount of its extent. When we measure, units stand in for rigid objects, and children's intuitions about spanning a magnitude with objects are employed as an entrée to eventually coming to conceptualize a measure as a ratio of the magnitude of the to-be-spanned distance (or area, angle, or volume) and the magnitude of a unit. For example, a measure of 10 inches means that the distance spanned is ten times that spanned by one inch, and an area measure of 10 square inches means that the space covered is ten times that covered by a square with side length of 1 inch. The rigid object metaphor fractures a continuous attribute of space into discrete units and, in doing so, creates a quantity—a measured length, area, volume, or angle. Quantities allow us to compare attributes both additively, as in *how much more or less*, and multiplicatively, as in *how many times more or less*.

The second grounding metaphor relies on experiences of space as one moves continuously through it. For example, a length is a path, and a measure is a destination along that path. Thinking of length dynamically creates opportunities to expand and literally grasp other ideas in arithmetic. Thinking of length as a distance traveled makes fractional quantities tangible and meaningful (Lehrer & Pfaff, 2011). For example, children who think about fractions only as parts of a whole tend to have difficulty interpreting $\frac{7}{3}$ (Thompson & Saldanha, 2003). But $\frac{7}{3}$ meter is simply a distance traveled by stepping $\frac{1}{3}$ meter seven times. The travel/motion metaphor is also helpful for addressing a common student misinterpretation of measuring, one that we call boundary filling. Boundary filling involves perceiving measuring simply as placing units within boundaries of the attribute to be measured, such as length, and then counting the units (Lehrer, 2003). Students who view measures as collections of units usually do not conceive of length as distance traveled and often have difficulties making sense of changes in origin (zero point) or accumulations of units that involve fractional quantities. For example, they may count a distance of $3\frac{1}{2}$ units as $4\frac{1}{2}$ units because the part-unit occurs within the span of what, from the perspective of counting units, would be the fourth unit (Lehrer et al., 1998).

Relations between angle, area, and volume can also be elaborated through metaphors of travel and motion (Lehrer & Slovin, 2014). For example, angle can be construed as amount of turn as one rotates around a point (Clements, Battista, Sarama, & Swaminathan, 1996; Lehrer, Randle, & Sancilio, 1989; Mitchelmore & White, 2000). One can think of an area as generated dynamically when one length moves through another at some non-zero angle. A rectangular area is generated in this manner by one length moving at a right angle through another length (Kobiela & Lehrer, 2019; Kobiela, Lehrer, & Pfaff, 2010). A rectangular volume is generated by moving a rectangular area through a length at a right angle in the third dimension (Kobiela & Lehrer, 2019). Images of stick and motion are complementary; thinking of measured space as generated dynamically and continuously complements the dissection of space by units that is promoted by the rigid stick metaphor. In the chapters that follow, we will illustrate how the rigid stick and continuous motion foundational metaphors can ground students' initial and subsequent activity as they learn to reason about measurement in increasingly complex ways.

Our emphasis is on long-term development. In kindergarten, a child's initial encounter with measure often involves either directly comparing attributes of objects to order them on that attribute or comparing representations (such as strips of paper) that are understood to stand in for an attribute of the objects. These initial forms of qualitative comparison undergird increasingly more powerful forms of quantitative comparison during succeeding years of schooling, so that *more than* and *less than* are eventually subsumed by questions about *how much more* and *how much less*, or by *how many times more* or *how many times less*.

Our emphasis on development of understanding is guided by over three decades of collaborative research that we have conducted with elementary school teachers who steward the education of children from diverse economic and social circumstances (e.g., Lehrer, 2003; Lehrer, Jaslow, & Curtis, 2003; Lehrer, Jacobson, et al., 1998; Lehrer et al., 1998; Lehrer & Kobiela, 2019; Lehrer & Schauble, 2019; Manz, Lehrer, & Schauble, 2020). With teacher partners, we have explored new forms of instruction, described later in this volume, that were designed to support children's learning of particular ideas in measure. We have conducted programs of research to document how children's thinking about measure tends to grow under these conditions of supportive instruction—this information is regularly fed back to teachers and used to inform ongoing revisions of the program. Our work with teachers has addressed conceptions of measures of length, area, volume, and angle and, for extensions of spatial measure, to thinking about fractions and arithmetic operations with fractions, such as sums, products, differences, and quotients. We have also conducted extensive programs of research into students' understanding of measure in the context of science, especially in the domain of field ecology (Lehrer & Schauble, 2015a, 2019).

Instructional outcomes from this work include a series of classroom instructional units, intended for use with students from grades kindergarten through five, that support children's emerging theory of measure and, as mentioned, also connect instruction in measure to related operations in whole number arithmetic and rational number. The curriculum includes 12 units (each with multiple lessons) devoted to measure of length, 5 to angle, 5 to area, and 5 to volume (unit summaries are included in the chapters devoted to length, angle, area, and volume, and the complete units are available online at Routledge. com/9781032262734). Eight of the units include lessons devoted to employing measurement models to help students make sense of fractional quantities and arithmetic operations with fractions, such as multiplication. One unit is devoted to coordinating measures of length and angle to establish location (polar coordinates). All the units have undergone multiple rounds of tryout, research, and revision with public elementary school teachers and students. Each addresses a set of related concepts and is designed to encompass several days of instruction. Units begin with a brief exposition of foundational mathematics concepts that the unit addresses, formulated for practicing teachers. Classroom instructional activities are described in detail and annotated with typical kinds of student responses and benchmarks. Each of the lessons includes a formative assessment that supports teachers in diagnosing the range of student progress on these ideas in a particular class and a formative assessment conversation guide that suggests how teachers can leverage typical student responses to items to foster greater understanding. Many of the lessons include supplementary activities that were contributed by practicing teachers, and these supplementary activities are frequently updated as teachers use the lessons to assist student learning.

The units are complemented by the design and several cycles of test and revision of an assessment system that characterizes conceptual change in four dimensions of spatial measure (length, area, volume, angle). The assessment system also makes visible related changes in students' thinking about fractions and about arithmetic operations involving fractions. The assessment system includes traditional one-on-one assessment items and also formative assessments and in-class observational tools for teachers to use to characterize student thinking in ways that can guide and inform their ongoing instruction. The assessments can also be used to help parents and guardians see children's mathematical progress. We describe some examples of this assessment system, along with children's responses to components of the system, in Chapter 8.

Relation to Existing Scholarship

Our work bears some family resemblance to—but also differs in some key respects from—two recent scholarly efforts to bring measurement into an appropriately more prominent role in elementary mathematics instruction. Both of these approaches, Measure Up and the Children's Measurement Project, like our own, are focused on fostering children's conceptual apprehension of measurement rather than mere procedural competence, and all three share the overall goal of accounting for the long-term development of children's understanding of measurement.

Measure Up is an adaptation for U.S. classrooms of an elementary curriculum originally developed by Davydov (1975) and his colleagues. Like our work, Measure Up stands as a critical alternative to the exclusive number-and-counting-based approach that characterizes mathematics instruction in many U.S. schools. Davydov and his followers in the U.S. (e.g., Davydov & Tsvetkovich, 1991; Dougherty, 2008; Sophian, 2008; Venenciano & Dougherty, 2014) argued that the concepts that children are traditionally taught in U.S. schools are inadequate for orienting students to those mathematical structures that will play a central role throughout their subsequent education. For instance, it is well established that children whose first instruction focuses exclusively on natural numbers and counting objects often become confused when the game changes as they encounter fractions, where, all of a sudden, rather than being about counting objects, mathematics is now about measuring quantities (Venenciano & Dougherty, 2014). Suddenly, things children believed they have "learned" (e.g., a bigger number always indexes a larger quantity) no longer seem to apply. Davydov advocated that early mathematics instruction begin not with counting objects, but with comparing physical attributes of objects and collections and representing these comparisons symbolically with relational statements (Davydov & Tsvetkovich, 1991). These comparisons should spotlight the core concepts of unit and relations of equality and inequality. Counting should not be introduced by assigning tags to objects defined perceptually, but, rather, framed as the measurement of a quantity that consists of a defined unit of measurement used a whole number of times. Addition and subtraction are positioned as operations that may change these relations by, for example, making two initially unequal quantities equal by increasing or decreasing one of the quantities. Number is introduced only after children have received extensive experience with exploring and symbolizing relations of equality and inequality with continuous quantities, such as lengths, masses, volumes, and areas. Students are encouraged to explore and define mathematical structures such as associativity, commutativity, and inverse. What guides the overall sequence of instruction is the ideas, relations, and concepts that eventually lead most centrally and directly to algebra.

We acknowledge and appreciate the mathematical systematicity and consistency that this approach represents. We also share the conviction that it is critical to support children's reasoning about continuous quantities; as explained, the systematic exploitation of metaphors of travel and motion is one of our chief strategies for calling attention to this focus. At the same time, instruction and assessment (including national forms of assessment) in the U.S., where educational choices are made locally and are often based on historical practice rather than logical consistency from a disciplinary perspective, are committed to natural numbers and counting-based approaches as the appropriate initiation into mathematics for most students. Moreover, there is good evidence (for example, from programs like Cognitively Guided Instruction, as described by Carpenter, Fennema, & Franke, 1996), that many children benefit from instruction that capitalizes on their intuitions about increasing, decreasing, and comparing numbers of perceptually distinct objects.

We think of the two foundational metaphors that ground our instructional program as providing a bridge across the apparent disjuncture between counting and measuring as ways of initiating children into mathematics. The rigid stick metaphor provides a context for thinking about the prevalent model of multiplication as repeated addition. For example, $10 \times 2u$ means ten iterations, or "copies," of a composite unit. This is the "groups of" image of multiplication that pervades elementary education. The motion metaphor provides an alternative interpretation of multiplication as a dynamic stretch, where $10 \times 2u$ means that the product is ten times as long as $2u$. The expansion of the semantics of multiplication is even more apparent in two dimensions, where movement of a length through another length generates a new type of quantity—an area. Moreover, as we later elaborate, the two foundational metaphors provide opportunities for interpreting fractions simultaneously as operators and as quantities. To represent $\frac{9}{2}$ units, first partition (an operation) the unit into halves and then travel nine times as far as $\frac{1}{2}$ unit. Hence, something continuous can be rendered as discrete, and the attributes so rendered can be compared in terms of quantity. Moreover, quantities can be fractional because there are stopping points anywhere on the continuum, so the logic and need for increasingly fine partitions is established. Thus, as children explore the relations between these two ways of interpreting units and partitioning, they also receive early experiences with splitting (Confrey, 1995), accumulating, and comparing fractional quantities.

In the most general terms, Measure Up, with its early emphasis on expressing relationships of equality and inequality, is oriented toward a general emphasis on algebraic thinking as the presumed mathematical core and endpoint of development. Although we share its emphasis on the centrality of continuous quantity and on the importance of representing mathematical relationships symbolically, nonetheless we cultivate measure primarily in the context of supporting children's thinking about space. Algebraic and spatial reasoning are both mathematically justifiable cores, but space has arguably been underexplored as an orienting endpoint to elementary education, in spite of the fact that its role in many people's lives (e.g., in STEM and design professions) may be at least as important as that of algebraic reasoning. Because we foreground spatial reasoning, we lend more instructional emphasis to some concepts that are not particularly highlighted in Measure Up, such as the construction of units, properties of units, and generation of different dimensions of space. Moreover, as we have described, we capitalize on children's actions within and on the world as a source of intuitions that can support the early mathematization of spatial reasoning.

Although these are differences in starting point and relative emphasis, it is important to acknowledge that they are not mutually exclusive. Spatial reasoning can support algebraic reasoning (Boester & Lehrer, 2007), just as algebraic reasoning can bootstrap reasoning about space. On the other hand, the two projects are working from different commitments about the most useful starting points for children's thinking and, also, the endpoint toward which instruction is most directly oriented (e.g., algebraic vs. spatial thinking).

The second recent, comprehensive, scholarly approach to measurement that is most closely related to ours is the research conducted by Clements, Sarama, and colleagues in the Children's Measurement Project. Over the past two decades, the Children's Measurement Project has developed a wealth of information about the typical development of children's measurement knowledge. This project and ours share a number of common agenda items, such as (1) focusing the field's awareness on measurement as a key mathematical domain that has been relatively understudied; (2) conducting many careful studies of children's learning of measurement concepts; (3) designing and investigating instructional tasks to support student development; (4) articulating typical pathways of learning by formulating and testing student learning trajectories (Maloney, Confrey, & Nguyen, 2014; Simon, 1995); and (5) revising and ultimately, validating these hypothetical learning sequences with psychometrically based models of assessment data from tasks given to groups of children across a range of elementary grades (Barrett et al., 2012, 2017; Clements, Barrett, & Sarama, 2017; Clements et al., 2018; Cullen et al., 2018).

Moreover, like our approach to measurement, the Children's Measurement Project positions measure within the context of spatial thinking and emphasizes connections between geometry and measure within the subdomains of length, area, and volume (Clements et al., 2017; Sarama et al., 2021). Both projects favor instructional designs that progressively build measurement knowledge on a foundation of children's early intuitive and experiential knowledge. Moreover, in both cases, the perspective on development is primarily sociocultural in flavor in that the ultimate rate and direction of development are considered to be contingent on children's opportunities to learn and their prior learning histories. Because of the close connection between development and learning, we consider that a satisfactory account of the development of measurement knowledge necessarily incorporates a description of the key features of the instruction under which that development is fostered (Lehrer & Schauble, 2015b; Sarama et al., 2021). These key features include, among possible others, instructional tasks, tools, forms of representation, and the modes and means of disciplinary argument that are fostered within the classroom as normative practice (Lehrer, 2009).

However, there are also key differences between the two approaches, which follow from some difference in overarching goals, procedures for developing models of student thinking, and pedagogical emphasis. With respect to goals, the Children's Measurement Project is primarily oriented toward describing structures of children's thinking, including how concrete work with objects and materials can support subsequent development of more abstract and, eventually, mental operations. Consistent with this goal, Clements and colleagues started with a review of existing literature on measurement and from it provisionally sketched a coherent longitudinal story, a hypothetical developmental progression that consisted of a postulated sequence of learning that described prototypical benchmarks of qualitatively distinct forms of student thinking (Barrett et al., 2012; Barrett et al., 2017; Clements et al., 2017). Subsequent studies with children were used to revise and refine the initial sequence. The pedagogical approach that followed is consistent with the rigid stick metaphor described earlier and, accordingly, tends to foster a "cover and count" approach to measuring length, area, and volume.

The rigid stick metaphor supports early measurement performances, especially in length, but, when complemented by an emphasis on continuity and motion, prescribes earlier and more detailed attention to the generation of unit and, in particular, fractions of units. Integrating both metaphors supports a more mathematically general and extensible treatment, especially when coupled with bidirectional movement between operations on concrete materials and notation of the mathematical relations thereby represented. Incorporation of continuous metaphors also promotes the integration of measurement with whole number arithmetic and fractions and tends to forestall some common errors in grasping measurement, including the "border filling" misinterpretation that we described earlier.

Finally, the developmental progressions that we eventually formulated (one each for length, angle, area, volume, and rational number) tend to be relatively granular because they are constructed primarily to guide everyday teaching decisions and thus must be sufficiently fine-grained to pick up differences in children's strategies that are consequential for instruction and assessment. Our sequences or trajectories of learning also presume some pedagogical commitments that are consequential for what and how students learn. Consistent with our emphasis on a *theory* of measure, we repeatedly encourage students to notice and debate what it means to measure (Cobb, Boufi, McClain, & Whitenack, 1997). For instance, carefully chosen, nonstandard units can provoke discussion about attributes (what property of a cardboard footprint or a paperclip is relevant for measuring?), equivalence (Mary's unit is two times as long as Frank's. The table is 2 Frank-units long. How many Mary-units are required to span the table?), and relations between the unit and the length of the object being measured (Frank's unit is half as long as the table. The table is two times as long as Frank's unit). Classroom discussions about these matters are conceptually productive but presume histories in which mathematical argument has become a normative practice. Understanding unit entails exploring relations between different units and the value of the resulting measure, partitioning units, combining units and partitions, apprehending the multiplicative relationship between units and partitions (e.g., Thompson, 2011), and developing mathematical representations of these relationships. Throughout, the use of unifying travel and motion metaphors encourages children to recruit their own bodily experiences to make sense of these ideas and to reflect about "big" ideas and procedures across the realms of length, area, volume, and angle. (In addition to Lakoff & Núñez, 2000, see also Elrich, Levine, & Goldin-Meadow, 2006; Goldin-Meadow, Levine, & Jacobs, 2014; and Levine, Goldin-Meadow, Carlson, & Hermani-Lopez, 2018 for other approaches to the metaphorical and embodied nature of mathematical concepts.)

Organization of the Book

This volume is organized to communicate typical patterns of children's conceptual development, starting with a chapter (Chapter 2) that describes the research and development process that guided the work, as well as the contexts in which it was conducted. What follows are chapters devoted to measurement in each of the following subdomains: length (Chapter 3), area (Chapter 4), volume (Chapter 5), and angle (Chapter 6) measure. For each aspect of measure, we first describe the network of core concepts that collectively contributes to understanding measure in that realm. Then we describe conceptual pivots—forms of activity and instruction—that have been extensively used in participating classrooms and that reliably provoke development and articulation of these networks of knowledge, as documented by supporting research that we and others have conducted. For each subdomain of measure, we extend measurement to rational number by describing how core concepts of rational number can be supported by conceptions and practices of measure. We also describe how measure and geometry can extend children's conceptions of arithmetic operations. Chapter 7 describes how experiences with measure can be used to support students' learning about fractions and rational number. Chapter 8 explains the three forms of research that contributed toward tracking change in student learning in this project and provides examples of each. Typical forms of student thinking are described and structured as constructs (Wilson, 2005) so that student responses to assessment items can be used to diagnose particular ways of thinking about measure. In addition, assessment items are designed so that post-assessment conversations with students can encourage them to develop new knowledge, rather than simply recounting what they already have learned. We also describe how student thinking manifested during classroom instruction can be related to student responses to end-of-year tests to produce more robust and useful accountability assessment. Chapter 9 details the teacher professional development that has accompanied this work and documents some of the changes in teachers' thinking and performance that occurred over the course of our extended collaborations in diverse public school districts. Finally, Chapter 10 illustrates the role of measure in science instruction, with a special emphasis on students' participation in constructing, critiquing, revising, and interpreting measures to address their own questions about growth and ecology.

References

American Physical Society. (2006). Eratosthenes measures the earth. *APS Physics News*, *15*(6). https://www.aps.org/publications/apsnews/200606/history.cfm

Barrett, J., Cullen, C., Behnke, D., & Klanderman, D. (2017). *A pleasure to measure: Tasks for teaching measurement in the elementary grades*. Reston, VA: National Council of Teachers of Mathematics.

Barrett, J. E., Sarama, J., Clements, D. H., Cullen, C., McCool, J., Witkowski-Rumsey, C. ... Klanderman, D. (2012). Evaluating and improving a learning trajectory for linear measurement in elementary grades 2 and 3: A longitudinal study. *Mathematical Thinking and Learning*, *14*(1), 26–54. doi: 10.1080/10986065.2012.625075.

Barth-Cohen, L. A., & Wittmann, M. C. (2016, June). Expanding coordination class theory to capture conceptual learning in a classroom environment. *Proceedings of the 12th International Conference of the Learning Sciences, ICLS2016*, Singapore.

Boester, T., & Lehrer, R. (2007). Visualizing algebraic reasoning. In J. Kaput, D. W. Carraher, & M. Blanton (Eds.), *Algebra in the early grades* (pp. 211–234). Mahwah, NJ: Lawrence Erlbaum Associates.

Carpenter, T. P., Fennema, E., & Franke, M. L. (1996). Cognitively guided instruction: A knowledge base for reform in primary mathematics instruction. *The Elementary School Journal*, *97*(1), 3–20.

Cartwright, N. L. (1999). *The dappled world: A study of the boundaries of science*. Cambridge: Cambridge University Press.

Clements, D. H., Barrett, J. E., & Sarama, J. (2017). *Children's measurement: A longitudinal study of children's knowledge and learning of length, area, and volume*. Journal for Research in Mathematics Education Monograph Series (Vol. 26), Reston, VA.

Clements, D. H., Battista, M. T., Sarama, J., & Swaminathan, S. (1996). Development of turn and turn measurement concepts in a computer-based instructional unit. *Educational Studies in Mathematics*, *30*, 313–337.

Clements, D. H., Sarama, J., Van Dine, D. W., Barrett, J. E., Cullen, C. J., Hudyma, A. ... Eames, C. L. (2018). Evaluation of three interventions teaching area measurement as spatial structuring to young children. *The Journal of Mathematical Behavior*, *50*, 23–41.

Cobb, P., Boufi, A., McClain, K., & Whitenack, J. (1997). Reflective discourse and collective reflection. *Journal for Research in Mathematics Education, 28*(3), 258–277.

Confrey, J. (1995). Student voice in examining "splitting" as an approach to ratio, proportions, and fractions. In L. Meira, & D. Carraher (Eds.), *Proceedings of the 19th international conference for the psychology of mathematics education* (Vol. 1, pp. 3–29). Recife, Brazil: Universidad Federal de Pernambuco.

Crosby, A. W. (1997). *The measure of reality.* Cambridge: Cambridge University Press.

Cullen, A. L., Eames, C. L., Cullen, C. J., Barrett, J. E., Sarama, J., Clements, D. H. ... Van Dine, D. W. (2018). Effects of three interventions on children's spatial structuring and coordination of area units. *Journal for Research in Mathematics Education, 40*(5), 533–574.

Davydov, V. V. (1975). The psychological characteristics of the "prenumerical" period of mathematics instruction. In L. P. Steffe (Ed.), *Children's capacity for learning mathematics* (pp. 109–205). Chicago, IL: University of Chicago.

Davydov, V. V., & Tsvetkovich, Z. H. (1991). The object sources of the concept of fractions. In L. P. Steffe (Ed.), *Psychological abilities of primary school children in learning mathematics* (pp. 86–147). Reston, VA: National Council of Teachers of Mathematics.

diSessa, A. (1988). Knowledge in pieces. In G. Forman, & P. Pufall (Eds.), *Constructivism in the computer age* (pp. 49–70). Mahwah, NJ: Lawrence Erlbaum Associates.

diSessa, A. A., & Sherin, B. L. (1998). What changes in conceptual change? *International Journal of Science Education, 20*(10), 1155–1191.

Dougherty, B. (2008). Measure up: A quantitative view of early algebra. In J. J. Kaput, D. W. Carraher, & M. L. Blanton (Eds.), *Algebra in the early grades* (pp. 389–412). New York: Erlbaum.

Elrich, S. B., Levine, S. C., & Goldin-Meadow, S. (2006). The importance of gesture in children's spatial reasoning. *Developmental Psychology, 42*(6), 1259–1268.

Goldin-Meadow, S., Levine, S. C., & Jacobs, S. (2014). Gesture's role in early arithmetic. In L. D. Edwards, F. Ferrara, & D. Moore-Russo (Eds.), *Emerging perspectives in gesture, embodiment, and mathematics.* Charlotte, NC: Information Age Publishing.

Kobiela, M., & Lehrer, R. (2019). Supporting dynamic conceptions of area and its measure. *Mathematical Thinking and Learning, 21*(3), 178–206.

Kobiela, M., Lehrer, R., & Pfaff, E. (2010, May). *Students' developing conceptions of area via partitioning and sweeping. Paper presented at the Annual Meeting of the American Educational Research Association.* Denver, CO.

Lakoff, G., & Núñez, R. E. (2000). *Where mathematics comes from: How the embodied mind brings mathematics into being.* New York: Basic Books.

Lehrer, R. (2003). Developing understanding of measurement. In J. Kilpatrick, W. G. Martin, & D. E. Schifter (Eds.), *A research companion to principles and standards for school mathematics* (pp. 179–192). Reston, VA: National Council of Teachers of Mathematics.

Lehrer, R. (2009). Designing to develop disciplinary dispositions: Modeling natural systems. *American Psychologist, 64*(6), 759–771.

Lehrer, R., Jacobson, C., Thoyre, G., Kemeny, V., Strom, D., Horvath, J. ... Koehler, M. (1998). Developing understanding of geometry and space in the primary grades. In R. Lehrer, & D. Chazan (Eds.), *Designing learning environments for developing understanding of geometry and space* (pp. 169–200). Mahwah, NJ: Lawrence Erlbaum Associates.

Lehrer, R., Jaslow, L., & Curtis, C. L. (2003). Developing understanding of measurement in the elementary grades. In D. H. Clements, & G. Bright (Eds.), *Learning and teaching measurement. 2003 yearbook* (pp. 100–121). Reston, VA: National Council of Teachers of Mathematics.

Lehrer, R., Jenkins, M., & Osana, H. (1998). Longitudinal study of children's reasoning about space and geometry. In R. Lehrer, & D. Chazan (Eds.), *Designing learning environments for developing understanding of geometry and space* (pp. 137–167). Mahwah, NJ: Lawrence Erlbaum Associates.

Lehrer, R., Kim, M. J., & Jones, S. (2011). Developing conceptions of statistics by designing measures of distribution. *International Journal on Mathematics Education (ZDM), 43*(5), 723–736.

Lehrer, R., & Kobiela, M. (2019, April). Supporting dynamic conceptions of multiplication. *Paper presented in symposium, Multiplicative models of situations across disciplines, at the Annual Meeting of the American Educational Research Association,* Toronto, Ontario, Canada.

Lehrer, R., & Pfaff, E. (2011). Designing a learning ecology to support the development of rational number: Blending motion and unit partitioning of length measures. In Y. Dai (Ed.), *Design research on learning and thinking in educational settings: Enhancing intellectual growth and functioning* (pp. 131–160). New York: Routledge.

Lehrer, R., Randle, L., & Sancilio, L. (1989). Learning pre-proof geometry with logo. *Cognition and Instruction, 6,* 159–184.

Lehrer, R., & Schauble, L. (2015a). Developing scientific reasoning: The role of epistemic practices. In R. Lerner, L. S. Liben, & U. Mueller (Eds.), *Child psychology and developmental science 7th edition, volume 2: Cognitive processes* (pp. 671–714). Hoboken, NJ: John Wiley and Sons.

Lehrer, R., & Schauble, L. (2015b). Learning progressions: The whole world is NOT a stage. *Science Education, 99*(3), 432–437.

Lehrer, R., & Schauble, L. (2019). Learning to play the modeling game. In A. Upmeir zu Beizen, D. Kruger, & J. van Driel (Eds.), *Models and modeling in science education* (pp. 221–236). Cham, Switzerland: Springer.

Lehrer, R., & Slovin, H. (2014). *Developing essential understanding of geometry and measurement in grades 3–5.* Reston, VA: National Council of Teachers of Mathematics.

Levine, S. C., Goldin-Meadow, S., Carlson, M. T., & Hermani-Lopez, N. (2018). Mental transformation skill in young children: The role of concrete and abstract motor training. *Cognitive Science, 42,* 1207–1228.

Maloney, A. P., Confrey, J., & Nguyen, K. H. (Eds.) (2014). Learning trajectories and professional development. *Learning over time: Learning trajectories in mathematics education* (pp. 227–242). Charlotte, NC: Information Age Publishing.

Manz, E., Lehrer, R., & Schauble, L. (2020). Rethinking the classroom science investigation. *Journal for Research in Science Teaching, 57*(7), 1148–1174.

Mitchelmore, M. C., & White, P. (2000). Development of angle concepts by progressive abstracton and generalization. *Educational Studies in Mathematics, 41,* 209–238.

Piaget, J., Inhelder, B., & Szeminksa, A. (1960*), The child's conception of geometry* (E.A. Lunzer, trans). New York: W.W. Norton.

Sarama, J., Clements, D. H., Barrett, J. E., Cullen, C. J., Hudyma, A., & Vanegas, Y. (2021). Length measurement in the early years: Teaching and learning with learning trajectories. *Mathematical Thinking and Learning, 6*(2), 81–89.

Simon, M. A. (1995). Reconstructing mathematics pedagogy from a constructivist perspective. *Journal for Research in Mathematics Education, 26*(2), 114–145.

Smith, C. L., Wiser, M., Anderson, C. W., & Krajcik, J. (2006). Implications of research on children's learning for standards and assessment: A proposed learning progression for matter and the atomic-molecular theory. *Measurement: Interdisciplinary Research and Perspectives, 4*(1–2), 1–98.

Smith, J. P. III, Males, L., Dietiker, L. C., Lee, K., & Mosier, A. (2013). Curricular treatments of length measurement in the United States: Do they address known learning challenges? *Cognition and Instruction, 31*(4), 388–433.

Smith, J. P. III, Males, L., & Gonulates, F. (2016). Conceptual limitations in curricular presentations of area measurement: One nation's challenges. *Mathematical Thinking and Learning, 18*(4), 239–270.

Sophian, C. (2008). *The origins of mathematical knowledge in childhood.* New York: Routledge.

Stephan, M., & Clements, D. H. (2003). Linear and area measurement in prekindergarten to grade 2. In D. H. Clements, & G. Bright (Eds.), *Learning and teaching measurement, 2003 yearbook of the National Council of Teachers of Mathematics* (pp. 63–76). Reston, VA: National Council of Teachers of Mathematics.

Thompson, P. W. (2011). Quantitative reasoning and mathematical modeling. In L. L. Hatfield, S. Chamberlain, & S. Belbase (Eds.), *New perspectives and directions for collaborative research in mathematics education. WISDOMe monographs* (Vol. 1, pp. 33–57). Laramie, WY: University of Wyoming. http://bit.ly/14w0flA

Thompson, P. W., Carlson, M. P., Byerley, C., & Hatfield, N. (2014). Schemes for thinking with magnitudes: A hypothesis about foundational reasoning abilities in algebra. In K. C. Moore, L. P. Steffe, & L. L. Hatfield (Eds.), *Epistemic algebra students: Emerging models of students' algebraic knowing. WISDOMe monographs* (Vol. 4, pp. 1–24). Laramie, WY: University of Wyoming. http://bit.ly/1aNquwz.

Thompson, P. W., & Saldanha, L. A. (2003). Fractions and multiplicative reasoning. In J. Kilpatrick, G. Martin, & D. Schifter (Eds.), *A research companion to principles and standards for school mathematics* (pp. 95–114). Reston, VA: National Council of Teachers of Mathematics.

Thompson, T., & Preston, R. V. (2004). Measurement in the middle grades: Insights from NAEP and TIMMS. *Mathematics Teaching in the Middle School, 9*(9), 514–519.

Van Fraassen, B. C. (2012). Modeling and measurement: The criterion of empirical grounding. *Philosophy of Science, 79,* 773–784.

Venenciano, L., & Dougherty, B. (2014). Assessing priorities for elementary school mathematics. *For the Learning of Mathematics, 34*(1), 18–24.

Wagner, J. F. (2006). Transfer in pieces. *Cognition and Instruction, 24*(1), 1–71.

Wilson, M. (2005). *Constructing measures: An item response modeling approach.* New York: Taylor & Francis.

2

THE CONTEXT, GOALS, AND DESIGN OF THE RESEARCH

The education innovation and research that we describe were pursued within an ongoing collaboration between the authors and teachers in two elementary schools, both located in the same district in the Midsouth. The collaboration was initiated in a development phase around 2010, although even at that point the work built substantially on earlier research in measurement and geometry conducted by Lehrer in the Midwest (Lehrer, 2003; Lehrer, Jaslow, & Curtis, 2003; Lehrer & Pritchard, 2002; Lehrer, Schauble, Strom, & Pligge, 2001). The development phase of the project extended from 2012 to 2016 with staff at Mallard Elementary School who taught in kindergarten through grade five. In 2016 the focus of the project shifted to Sleeve Elementary School, where we initiated a tryout of the mostly stabilized design with teachers and their students from grades K–5. (Both school names are pseudonyms.)

Research Sites and Participants

Both Mallard and Sleeve Elementary Schools are located within the fourth most populous district in Arkansas. However, the high population is due to the wide geographic area that the district covers; its population is not dense. The district includes a small city, suburbs, and a wide swath of rural areas. The median household income is about $50,000, and 16.7% of the population live below the poverty line. Nearly a quarter of the residents were born outside of the U.S.

Mallard Elementary School is among the top 1% most diverse public schools in the state. Children of Hispanic heritage comprise 37% of the students; 27% are Marshallese and 29% are white. At Sleeve Elementary School, 83% of the students are white and 14% are Hispanic. The parents of students in the district work in a variety of occupations, especially manufacturing, meat packing,

DOI: 10.4324/9781003287476-2

trucking, and retail sales. In addition, some parents occupy middle management positions in a large national chain store that is headquartered nearby.

The participating teachers were all regularly assigned public school elementary teachers, with widely varying experience (from novices to experienced practitioners with over a dozen years of teaching). Except for the two mathematics coordinators in the participating schools, most of the teachers had no special emphasis in mathematics education in their training. However, the district where the schools are located had an existing affiliation with researchers representing Cognitively Guided Instruction (CGI), an elementary mathematics instructional initiative that is well aligned with our own emphasis on understanding student thinking and foregrounding student reasoning and argumentation during instruction (Carpenter, Fennema, & Franke, 1996; Carpenter, Fennema, Franke, Levi, & Empson, 2015; Carpenter, Franke, Johnson, Turrow, & Wager, 2016). The district mathematics coordinators had received professional training in CGI and had worked for several years to establish that approach in each school. This history provided an important head start for the current project, because participating teachers were at least aware of the importance of (and were variably skilled at) basing instruction on their understanding of evolving student thinking.

Administrators at both Mallard and Sleeve agreed that students enrolled in participating classrooms would be preferentially assigned the following year, insofar as possible, to classrooms at the next grade level who were also participating. This arrangement was intended to support the longitudinal design of the curriculum and related studies of student learning, but it was not fully implemented at either school, as classrooms typically mixed continuing participants with newcomers. As a result, some of the teachers, especially in the older elementary grades, often had to step back to address foundational concepts that were unfamiliar to the newcomers in their classes. The tryout at Sleeve leveraged more stabilized curricula and digital tools but was disrupted by the effects of the global pandemic.

Design Research: Engineering Learning to Support Its Study

The development and research efforts pursued in these settings belong to a genre that is becoming increasingly familiar to education researchers as *design research*, a research orientation that involves designing target forms of learning and then systematically studying them as they emerge within the context where they are being supported (Bakker, 2019; Cobb, Confrey, diSessa, Lehrer, & Schauble, 2003). The designed context is subject to repeated iterations of test and revision, a process that produces information about systematic variations of the design and thereby helps researchers determine which features are critical to learning and how they operate within the context. Design research, therefore, is the coordinated pursuit of two related goals: (1) designing and supporting particular (usually innovative) forms of learning and (2) systematically studying those forms of learning within the designed context.

Design studies are not intended to confirm or disconfirm comprehensive learning theories like activity theory or sociocultural theory, nor, at the other end of the spectrum, are they mere attempts to tinker with contexts to see "what works." Instead, they are "crucibles for building and testing theory" about learning (Cobb et al., 2003, p. 9), but focused at a midlevel of generality with a grain of analysis sufficient to identify domain-specific learning processes. The purpose is to understand whether and why the design in question works and how it can be successfully adapted to new contexts. Characteristically, design studies are testbeds for innovation rather than studies of learning under usual conditions. Because of their relatively constrained focus, Cobb and colleagues refer to them as "humble theories," but nonetheless their goal is to advance theory, not simply to improve instruction in the research site. In sum, design research characterizes patterns in students' learning of specific domain concepts and forms of reasoning and is aimed toward developing a principled account of what it takes to support their development.

At Mallard and Sleeve, we sought to engineer a learning ecology to support students' deep and principled understanding of measure. The schools provided opportunities to study how learning evolves longitudinally (in this case, over the five years of elementary school), as well as to continually update our initial conjectures about student learning as we were informed by ongoing results of learning studies in our participating classrooms. In this manner, design research also helped refine our understanding of what it takes to bring these forms of learning into being. Contexts, of course, are complex and may be interrogated at different levels of analysis. We chose to document learning at the analytic levels of individual students, classrooms, and the professional teaching community. At each of these three levels, the research sought to identify the design elements that are most essential and then to characterize their role in the design. For instance, at the classroom level, design elements included the tasks or problems with which students were asked to engage, the kinds of discourse and norms of participation teachers sought to instantiate among their students, the tools and other material means on which students and teachers relied, and the teachers' work to orchestrate these elements in their classrooms. We turn now to providing a brief description of some of the essential features of the design.

The Educational Design

There were five essential elements of the educational design: (1) the professional development collaboration with participating teachers, who implemented instruction with students; (2) learning constructs that provided an analysis of and language of learning shared by all stakeholders; (3) a novel mathematical curriculum that simultaneously guided and described instruction and classroom activity; (4) an assessment system that integrated formative and summative information about student progress; and (5) digital tools for administering, collecting, managing, displaying, and interpreting student data. These design elements were developed in several successive prototype versions that were iteratively revised on a continuing basis as feedback from their use was generated in participating classrooms. Thus, they were simultaneously tools for implementing the design and, ultimately, products of the research. Eventually, the constituents of the learning ecology were stabilized into recurring routines and supporting materials for professional development; published and website versions of curriculum, constructs, and learning tools; formative and summative assessment items and routines; and technological systems to administer and interpret student assessment at both the classroom and individual student levels.

Professional Development Collaboration with Participating Teachers

Table 2.1 displays the numbers of teachers at each grade level who worked on this research each year of the project, initially at Mallard Elementary School and subsequently at Sleeve Elementary School. Some of the participating teachers switched grades from year to year, and the first- and second-grade

TABLE 2.1 Numbers of Participating Teachers by Grade

	2012	2013	2014	2015	2016	2017	2018	2019	2020	2021
Pre-K										
Mallard	–	–	1	–	–	–	–	–	–	–
Kindergarten										
Mallard	2	3	3	3	3	1	–	–	–	–
Sleeve	–	–	–	–	–	–	2	2	2	–
First Grade										
Mallard	3	4	3	4	4	–	–	–	–	–
Sleeve	–	–	–	–	–	–	4	4	4	3
Second Grade										
Mallard	4	2	4	3	3	1	1	1	1	–
Sleeve	–	–	–	–	–	–	1	2	1	2
Third Grade										
Mallard	1	3	1	6	6	2	1	1	–	1
Sleeve	–	–	–	–	–	–	2	3	2	2
Fourth Grade										
Mallard	4	3	3	2	2	1	1	1	–	–
Sleeve	–	–	–	–	–	–	4	2	2	3
Fifth Grade										
Mallard	5	2	1	4	4	1	2	2	1	1
Sleeve	–	–	–	–	–	–	1	1	1	1
Math Coordinator										
Mallard	1	1	1	1	1	–	–	–	–	–
Sleeve	–	–	–	–	–	–	1	1	1	1
Totals	20	18	17	23	23	7	20	20	15	14

teachers in both schools regularly "looped," teaching first grade one year and then moving with their students to the second grade. The development phase of the project occurred primarily at Mallard, where we piloted and revised most aspects of the design. After a transition year in 2017, we then turned our attention to Sleeve, where we implemented a tryout with the mostly stabilized design. However, some revisions to aspects of the design continued while we were at Sleeve, along with additional development of content to capitalize explicitly on intersections between measure and fractions/rational number. Although we no longer engaged in intensive data collection at Mallard after 2016, a few Mallard teachers continued as research participants. They played an important role in reviewing curriculum drafts and professional development materials and assisting with continuing professional development at the Sleeve site. Once we shifted to Sleeve, the curriculum and assessment materials continued to be used at the original Mallard site, with participating teachers there taking a leadership role in the implementation and coaching of newcomers in this approach.

The intervention depended on direct implementation by regular practicing teachers. Therefore, an essential part of the overall design was professional development to help teachers understand how student knowledge about measurement develops and how that knowledge can best be supported. Chapter 9 provides further documentation about teacher learning. Here, we briefly outline the major forms of professional development that occurred.

Because the researchers did not live in the state where the participating district was located, some of the continuing exchange occurred through email and internet-based platforms, especially during the 2020–2021 pandemic year. During the other years, researchers traveled to the participating schools each month for regularly scheduled meetings. These meetings typically lasted for two to three days on each monthly trip and for a week during each summer. The content and format of the meetings varied in response to teacher need but typically included two major components: (1) work with the whole group on a broad variety of professional development activities and (2) embedded teaching in the classrooms of participating teachers. Whole-group work included a study of relevant mathematical ideas (sometimes led by consulting mathematicians, especially during the summer sessions), discussions of episodes of student thinking and performance that were recorded from teachers' classrooms, application and refinement of scoring systems to assess student work, review of drafts of curricular and assessment materials, and interaction with/feedback on pilot versions of technological systems and tools being created or adapted for the project. Embedded teaching was a format that the group adopted during the final year at Mallard and subsequently used regularly when the professional development shifted primarily to Sleeve. Our implementation was most closely aligned with math labs, as described by Kazemi et al. (2018), and with Japanese lesson study, as described by Lewis and Perry (2017).

Teachers first identify a teaching concept or activity—usually one featured in one of our project curricular units—that they feel poses new teaching challenges. Teachers then collectively plan a lesson to anticipate how they will elicit and support students' thinking about a focal concept. Classroom implementation is led by one or more teachers, and as the lesson unfolds, participating teachers conduct brief diagnostic conversations with a range of students to develop a sense of how they are thinking about the concepts being taught and about how intended supports for learning are functioning. Occasionally, a team member may call a "time out," a brief pause in instruction during which the observing teacher(s) or the lead teacher(s) raises a question, debates the interpretation of student work and conversation, or proposes a minor detour in the instructional plan. Once the class has ended, the session is debriefed by the team, who share what they observed and, as a group, discuss potential productive revisions or additions, as well as productive next steps in instruction. Often, the revisions or additions to the lesson are subsequently implemented in other classrooms, followed by another round of shared feedback and (potentially) further refinement of the lesson. Typically, several of these embedded lessons were conducted during each of the researchers' monthly visits. As we will illustrate in Chapter 9, the embedded lessons were a powerful force for building teacher community and knowledge, especially when participants included a cross-grade group of teachers. In these cross-grade groups, teachers debated the implications of earlier learning for learning in later grades. They reported that they rarely had opportunities to consider this longitudinal perspective on student achievement. The conversations also provided feedback to help researchers refine and improve the project curriculum units on an ongoing basis. During the COVID-19 pandemic, travel between states was discouraged, so the embedded teaching sessions and subsequent planning were led for that period by Sleeve's instructional facilitator.

Learning Constructs

A second essential part of the design was the generation and refinement of learning constructs (Wilson, 2005), documents that summarize typical progressions and patterns of student thinking within each of the major domains of interest—in this case, the measurement of length, angle, area, and volume and rational number. Constructs are summary portraits of the development of children's thinking within the specified content domain, and they are pivotal in that all elements of the design—curriculum, assessment, and professional development—are formulated to define learning progress in relation to the characterizations of student thinking that the constructs articulate. Constructs identify benchmarks that describe and order (from less to more sophisticated) qualitative shifts in thinking about a content domain. For example, there are six major benchmarks in children's understanding of the measurement of length, starting with the kind of understanding that kindergartners are likely to display and then continuing through ways of thinking that are increasingly mathematically challenging. (The complete learning construct for length, which describes these benchmarks in finer-grained detail, is discussed in Chapter 3.) Each of these major benchmarks includes several sublevels, which are also delineated in the learning construct.

- Assessing length by directly comparing lengths
- Explaining properties of units and how properties ground accumulation (count) of units
- Iterating units and symbolizing length as distance traveled
- Equipartitioning units by 2 and symbolizing partitioned units on a measurement scale, including zero point
- Partitioning units by 3 and composing partitions of 2 and 3
- Generalizing relationships among magnitudes

In addition to playing a role in clarifying learning goals for students, constructs are also important tools for supporting professional development, because they offer teachers a common lens for understanding the development of mathematical ideas and ways of assessing student progress in understanding those ideas. The learning constructs for each of the domains represented in the measurement research are discussed in the relevant chapters devoted to measurement of length (Chapter 3), angle (Chapter 4), area (Chapter 5), and volume (Chapter 6) and rational number (Chapter 7). Chapter 9 describes how teachers use constructs to develop instructional practices responsive to student thinking.

Constructs are the theoretical lynchpins of the educational design in that they provide a common definition of learning progress that informs and guides the blueprint of all other elements. For instance, both curriculum and assessment are calibrated to the levels of student thinking that the constructs articulate, so instruction and assessment share a common language of description. They also give stakeholders a common picture of progress and how to recognize it, thereby increasing the likelihood that teachers, administrators, parents, and external accountability assessment professionals will agree on what is valued and how to see it. Moreover, those performances are described as increasingly sophisticated levels of thinking, not simply as correct or incorrect answers to tasks. They identify a range of student ways of thinking about target concepts and thereby help teachers differentiate and make sense of student responses that they might otherwise simply lump into a single "incorrect" bin. Thus, they are a more comprehensive and fine-grained way of diagnosing student performance than is provided by some assessment systems. Their granularity and content specificity help educators identify next appropriate instructional goals that are tailored for a range of students.

Articulating a learning construct involves identifying the resources that beginners are likely to bring to their early encounters with tasks in the relevant domain(s) and then characterizing benchmarks that represent key conceptual shifts that occur as knowledge and proficiency develop. Practically, this entails combing through existing relevant research and, in addition, conducting new, targeted learning studies to (1) define desired endpoints of performance as competencies, along with illustrative examples; (2) characterize conceptual and performance resources that students typically have that could be recruited for developing those ideas and skills; and (3) document intermediate ways of thinking at a level of specificity sufficient to guide teaching decisions. These intermediate ways of thinking are regarded not as Platonic stages of domain-general development, but rather as domain-specific ways of thinking that are presumed to be conditional and based on the presence and quality of forms of instruction that are designed to address and support them. Like all other elements of the instructional design, the learning constructs were iteratively revised throughout the duration of the project in response to feedback from learning studies and observations of students in the participating classrooms.

Supporting Curriculum Units

Guided by the learning constructs, researchers and teachers developed, tested, and iteratively revised the third essential element of the design: a comprehensive mathematics curriculum to guide teaching of the measurement of length, angle, area, and volume (altogether, approximately 135–150 hours of instruction across kindergarten through grade five) and, throughout, to delineate productive connections between measurement, arithmetic operations, and rational number.

The curriculum appears to devote more hours of instruction to measurement than students traditionally receive, but in fact the materials pay considerable attention to reinforcing connections to mathematics across the elementary curriculum. They therefore serve as additional contexts for working on related strands of mathematics, especially arithmetic operations (addition, subtraction, multiplication, and division) and rational number. For that reason, and because the materials focus on fundamental ideas about unit, accumulation, scale, and partitioning, the curriculum is considerably broader and more integrative than lessons focusing solely on measurement.

The original purpose for developing the curricular units was to guide instruction in the schools that are the contexts for the research, but the units are also intended to have prospective utility for other teachers, both within the district and beyond. To this end, curriculum to guide student learning at each grade is formatted to be educative for teachers (Davis & Krajcik, 2005). Each of the lessons in a unit provides a pithy explanation of key mathematical ideas and their importance. Each unit includes schematic conceptual diagrams to help teachers keep track of progress in students' mathematical development (e.g., see Figure 3.2 in Chapter 3 of this volume). (All the units are available for review or download on the website Routledge.com/9781032262734) In addition to student activities, units include teaching suggestions and coding/grading schemes for diagnosing and responding to differential student strategies and knowledge as they appear during curriculum-related activities. Examples and extensions of instructional approaches that are contributed by the participating teachers are featured in a "Teachers' Corner" section of each unit. Formative assessment routines and conversation guides help teachers characterize the progress of each student and allow them to view an evolving class profile with respect to the core concept(s) that the unit targets.

Across the curriculum, units share a focus on encouraging teachers to center instruction around students' discourse, with an emphasis on engaging the class in mathematical argument and justification at an appropriate level of sophistication. For example, kindergartners debate which of several pumpkins is the "largest," opening consideration of the range of attributes that might count as "large" and how to measure them. First graders deliberate about how to label a length that falls between one unit and two units. Is it one and a half or two and a half units long? Discussions like these tend to engage variable forms of children's intuitive knowledge, in turn provoking questions for the class to consider about how to decide and what counts as evidence. In these ways, the curricular design is intended to engage teachers and children early on in increasingly mathematical forms of thinking, such as definition, generalization, and proof (explaining why).

The process of curriculum development extended throughout the duration of the project. Initial and intermediate prototypes of units were taught in multiple classrooms, and careful observations of children's performance and forms of thinking during task tryouts were recorded and used to inform rounds of revision. Ongoing feedback from educators was systematically sought within and between scheduled group meetings. As the content of these curriculum units began to stabilize across the years of the project, researchers increasingly turned their attention to issues of format, experimenting with presentational variants to ensure that material would be optimally accessible, understandable, and useful to teachers.

Units were developed and revised grade by grade on a rolling basis. Early years of the project focused on generating material for the younger grades, with most of the initial emphasis on the measure of length and angle. As participating students "graduated" into the upper grades, curriculum design progressively shifted to focus on ideas taught in the third, fourth, and fifth grades, especially the measurement of area and volume. Most recently, curriculum development has emphasized maintaining a consistent and coherent focus on connecting measurement to arithmetic and rational number and fractions across all four domains of measure.

The units are intended to be taught sequentially, with later concepts building systematically on those taught in earlier grades. Hence, in the research, the expectation is that students in later grades have already constructed a solid understanding of concepts taught earlier. Therefore, if the curricular units are used in research contexts other than ours, grade-level judgments may need to be adjusted for populations or individual students who enter the sequence in the upper grades without having grappled with the fundamental underlying ideas taught in lower grades. Our experience has been that it is possible to catch students up without starting all the way back at the beginning lessons, but teachers need to take seriously the problem of how to establish key concepts like unit, travel, and iteration, even if those ideas are presented in abbreviated form with older learners. One cannot simply assume that students who have received typical instruction in measurement will sufficiently understand these underlying concepts.

Assessment System

An assessment system that is aligned with the learning constructs is the fourth essential element of the design. Often, teachers' formative judgments and formal assessment systems do not mesh well and therefore do not routinely inform each other. This is because they are not typically based on a common framework that articulates a coherent vision of what is being assessed (Wilson, 2005). In particular, the construction of high-stakes assessments is rarely guided by an underlying cognitive model based on any coherent theory of knowledge development. Most high-stakes assessments are based instead on psychometrically constructed models of student performance that rely primarily on measures of item difficulty, which are based, in turn, on the percentiles of students who respond correctly to assessment items. The primary purpose of assessments like these is to rank student (or school) performance in comparison to other reference groups, not to deliver usable information about how to diagnose or effectively support conceptual development, over either reasonably brief (e.g., within one academic year) or longer periods of time.

To address these and related problems, Dr. Mark Wilson and his colleagues at the University of California–Berkeley, created the Berkeley Evaluation and Assessment Research (BEAR) Assessment System, which provides the format for the assessment design used by our project. The BEAR Assessment System has been in development over the past 20 years to support teacher-managed, classroom-centered assessment of student performance from a learning progression perspective. This perspective means that assessments are built to align with cognitive theory about the typical development of concepts that undergird understanding of the target domain. A key commitment of this approach is that assessments should match the purpose and content of the instruction in which the assessments are embedded. For more detail on the BEAR Assessment System, see Wilson and Sloane (2000).

As the BEAR Assessment System prescribes, our learning constructs describe important variants and shifts in student thinking, thus articulating a theory of the development of student knowledge at a practical level of analysis that is sufficient to guide both instruction and the development of assessment. Each of these shifts, or benchmark levels, is accompanied by examples of student performances that are diagnostic of that level. For instance, Table 2.2 presents the first level of the construct for length measurement (Theory of Measure, Length), which describes conceptual milestones toward establishing relations between magnitudes of length by directly comparing them (Benchmark 1). The benchmark level, 1, is partitioned into multiple sublevels (e.g., ToML 1A, 1B). Each sublevel articulates (the middle column) and exemplifies

TABLE 2.2 First Level of the Learning Construct for the Development of Length

1	*Directly Comparing*	
Level	*Performances*	*Examples*
ToML 1E	Distinguish (e.g., equal, not equal) or order (e.g., greater, lesser) magnitudes of an attribute by direct comparison of representations. Symbolize comparisons established with written expression.	"Pumpkin A is taller than pumpkin C" (Student aligns paper strips that stand in for height and notices that A's strip is longer than C's strip). $H_A > H_C$ (Student interprets and selects appropriate pair of strips among several candidate pairs of pumpkin heights.)
ToML 1D	Distinguish (e.g., equal, not equal) or order (e.g., greater, lesser) magnitudes of an attribute by direct physical comparison.	"This book is taller than that one (Student aligns the books and compares)." (Paired comparisons tend to be more accessible to students.) "Johnny is tallest, and Sally is in the middle, and Jennifer is the shortest." (Multiple ordinal relations tend to be more difficult for very young children).
ToML 1C	Define the attribute being measured so that measures can be compared.	"Fat means how far it is around the caterpillar" (analogy to circumference of wrist). "Big means the one that weighs the most." "Height is straight from floor to the top of the pumpkin."
ToML 1B	Identify measurable attributes (qualities).	"We could find out how long the caterpillar is or how fat it is." "We can find the height of each pumpkin."
ToML 1A	Pose a question or make statements about a potentially measurable object of interest.	"How big is the pumpkin? "Which rocket flies the best?" "Which pumpkin is tallest?"

(the rightmost column) learning performances that are taken as evidence of ways of thinking, directly comparing typical of measurement novices—for example, students in kindergarten or first grade. Table 2.2 is intended to be read from the bottom (where the earliest forms of knowledge are described) to the top (at each succeeding sublevel, more complex forms of thinking are articulated). It is important to understand that the progress in thinking that the figure describes is not "development in general," but domain-specific cognitive development that is contingent on instruction. As the table illustrates, later-developing ideas are intended to acknowledge, build from, and capitalize on earlier, more intuitive forms of knowledge that are relevant to the domain. Accordingly, this is an approach that emphasizes the continuities between intuitive and instructed knowledge.

Assessment items are based on the learning constructs; they are designed to characterize student responses to assessment tasks with respect to the benchmark level(s) of thinking. Many of the assessment items are designed specifically to distinguish performances at more than one level, so they can differentiate more from less sophisticated levels of thinking about the same problem. Item scoring is based on learning construct levels and sublevels. Some of the items in the measurement assessment item bank are brief; others are based on complex task scenarios and require linked responses. Currently there are about 250 items in the project databank, although new items are always being added. Some are adaptations of items used in earlier research by project team members and other investigators; others were designed especially for this research. Most of the items have now been administered multiple times and their psychometric properties have been established. They have been used through multiple rounds of administration, scoring, and review; they have been compiled into tests to assess student progress on targeted ideas; and they have been scored, with the results used to inform educators and administrators.

Because instruction and assessment derive from a common framework, tasks conducted as part of regular classroom instruction provide an additional potential source of information that can be mined for summative assessment purposes. A novel development goal for the project is to explore how formative, classroom-based assessment can provide actionable information for improving classroom instruction and, at the same time, contribute to system accountability decisions while simultaneously addressing the demands of psychometric quality. Our partners at the University of California, Berkeley, are exploring new psychometric models that link information from multiple sources, including teachers' judgments about student classroom work, student responses to formative assessments, and more traditional summative evaluation items (Wilson, 2021). The purpose of including these new, classroom-based sources is to expand the information that informs accountability decisions, so that these decisions can capitalize on the increased density and timeliness of the judgments that teachers make on an everyday basis. At the same time, we are also seeking to increase the feedback frequency with which accountability information is returned to teachers, so that they can benchmark individual students and classes as instruction proceeds and adjust their instruction accordingly, rather than receiving assessment results only when the school year is over. In addition to basing both classroom-based and summative assessment on a common learning constructs framework, we have been inventing digital tools that support teachers as they record, characterize, and interpret summary displays of student performance at multiple levels: individual student, class, and grade level.

Digital Tools for Collecting, Displaying, and Interpreting Student Data

Two systems of digital tools are regularly used by participating researchers and teachers. The first is the Berkeley Assessment System Software (BASS), developed by colleagues at the University of California, Berkeley. BASS stores, analyzes, and reports data for several classroom-based learning studies, including ours, that work in partnership with the BEAR Assessment System.

BASS was built to support the storage, delivery, analysis, and display of results for assessments that are aligned with learning constructs. BASS has been in continuing development for over 15 years and has been adopted by educators and researchers who represent over a dozen independently funded research and development projects housed in university and research institution settings. To support our collaboration, the BEAR and Vanderbilt teams collaboratively designed several human-interaction additions to the features available in BASS. These include the incorporation of audio versions of instructions, including a Spanish version; video-based examples for young students; support for the use of directly manipulable screen tools like protractors and measurement units; specialized drawing tools for expressing responses; and other forms of action embedded into items, such as splitting or dragging. Items that employ these new capabilities were pilot tested and revised for user comprehensibility.

The second system of digital tools is the Teacher Observation Tools (TOTs), designed by the Vanderbilt team under the leadership of Professor Corey Brady. TOTs is the system that teachers use to record formative classroom information and to display the results in ways that guide instructional decision-making. One of the novel development goals for the project was to explore how formative, classroom-based assessment can provide actionable information for improving classroom instruction and, at the same time, contribute to system accountability decisions while addressing the demands of psychometric quality. In addition to providing direct feedback on instructional progress in the classroom, the TOTs tools also serve as the portal for delivering class-room-based data to the BASS system.

TOTs is a web-based toolkit, implemented on iPads, that supports teach-ers as they record, store, and display classroom-based evidence of student thinking in a variety of formats, including video or photos of classroom work and textual commentary. As teachers record this evidence during or after instruction, they use a built-in coding system to score each segment of evidence by associating it on a menu with the appropriate sublevel of the appropriate learning construct. Each of these indications of student thinking is subsequently considered to be an "item" in the BEAR model that inte-grates formative with summative assessment. TOTs reports data to BASS so that classroom data can be incorporated into the integrated assessment models that are being developed at the University of California, Berkeley. In addition, the classroom evidence can be displayed for teachers' interpretation to inform ongoing instruction. Evidence in TOTs may be in the form of recorded student talk, video, photos, strategies, work products, and directly observed student activity (e.g., the way a student iterates a measurement unit to measure a length).

To use TOTs effectively, teachers must learn to read and register examples of student thinking as they occur during classroom activity by associating the examples with the appropriate levels and sublevels of the relevant learning construct. Thus, teachers' professional development is key to the practicality of this kind of assessment, and, reciprocally, using the assessment serves as a productive context of professional development. Using both BASS and TOTs helps sharpen teachers' noticing as they seek to characterize key examples of student talk and activity in relation to the learning constructs. Because this information guides their upcoming instructional decisions, teachers learn to leverage their knowledge of student thinking to improve the quality of instruction, so that assessment becomes a vital component of instructional practice, guiding the shared norms and interpretations of the surrounding community, rather than simply serving as an external report card that arrives long after instruction is completed.

Figure 2.1 is a facsimile of the teacher recording portion of the TOTs toolkit. It displays a photo, presumably taken by a teacher during an ongoing class activity, along with a teacher textual comment. The teacher has selected the construct sublevel that she believes describes the student thinking that the photo illustrates. The recording system allows her to attribute this perfor-mance to either a single student or a group of students.

TOTs also features visualization tools that allow teachers to review student data, either from a single classroom or across the relevant grade level or school. Figure 2.2 is a facsimile of a dot plot of evidence that characterizes students' thinking about the measurement of length. Each of the labels at the top of the figure (1A, 1B, etc.) indexes a level and sublevel of the length learning construct.

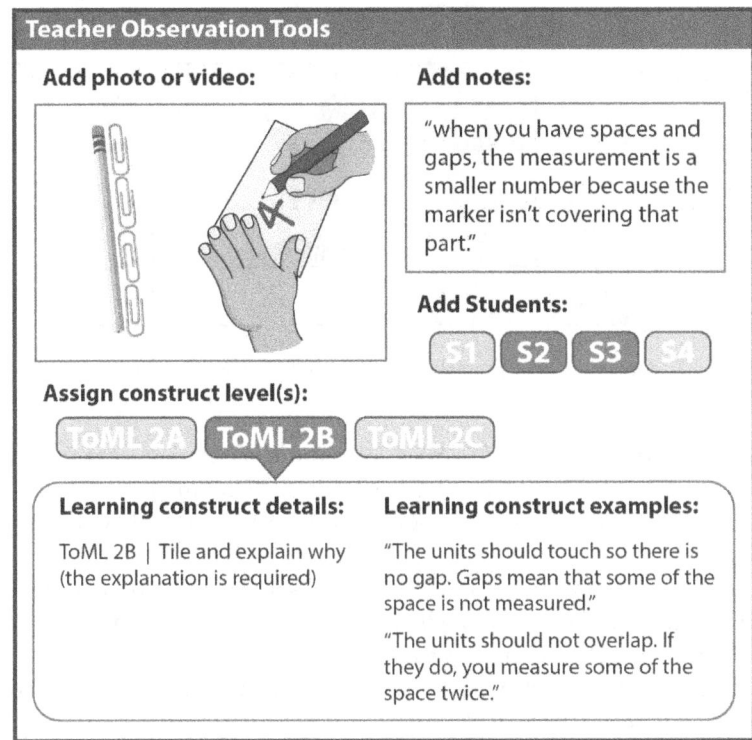

FIGURE 2.1 Teacher recording portion of TOTs

Dots correspond to observations. If a teacher clicks on a dot, the contents of the observation are revealed (in this case, information about the observation, as in Figure 2.1, is shown). This display provides a general picture of the class's current progress with respect to a given construct.

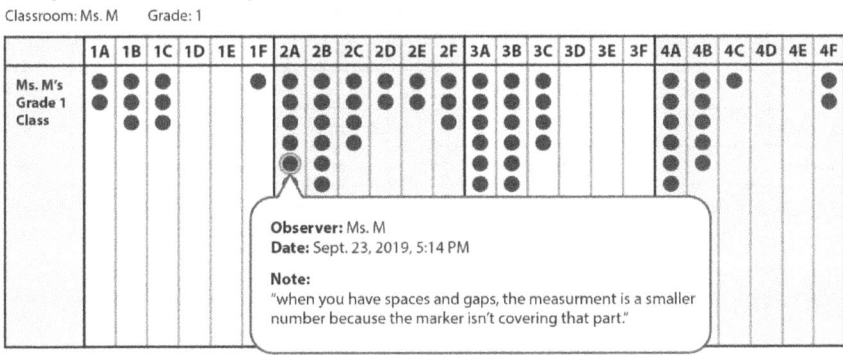

FIGURE 2.2 A dot chart that displays data at the sublevel of the construct

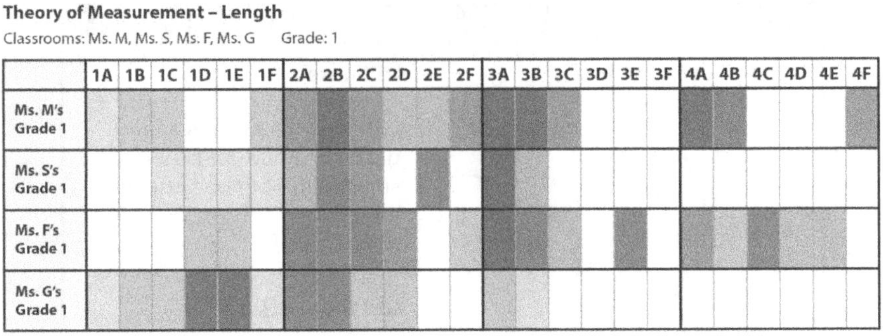

Theory of Measurement – Length

Classrooms: Ms. M, Ms. S, Ms. F, Ms. G   Grade: 1

	1A	1B	1C	1D	1E	1F	2A	2B	2C	2D	2E	2F	3A	3B	3C	3D	3E	3F	4A	4B	4C	4D	4E	4F
Ms. M's Grade 1																								
Ms. S's Grade 1																								
Ms. F's Grade 1																								
Ms. G's Grade 1																								

FIGURE 2.3   A construct heat map that displays data across classrooms

An alternative display of data uses color intensity to represent frequency of observation. As illustrated in Figure 2.3, this "heat map" view can summarize observations across classrooms, producing a school-wide view of progress.

A "star chart" view, shown in Figure 2.4, displays observations at sublevels of the relevant construct for individual students. This display makes it possible for teachers to develop a sense of the variability of student thinking in their class and to ensure that they are not basing their judgments on impressions from just a few select students.

The design of these alternative display modes was informed by teachers' preferences about the kind of information they wanted from their formative assessment efforts and how they wanted it displayed. As with other essential elements of the project's overall design, digital tools were iteratively revised,

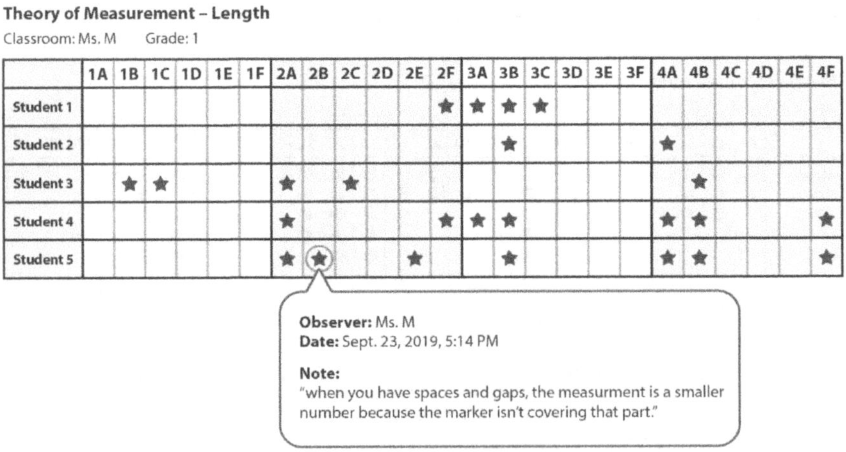

Theory of Measurement – Length

Classroom: Ms. M   Grade: 1

	1A	1B	1C	1D	1E	1F	2A	2B	2C	2D	2E	2F	3A	3B	3C	3D	3E	3F	4A	4B	4C	4D	4E	4F
Student 1							★	★	★	★														
Student 2									★										★					
Student 3		★	★				★		★											★				
Student 4							★			★	★	★							★	★				★
Student 5							★	⊛		★			★						★	★				★

> **Observer:** Ms. M
> **Date:** Sept. 23, 2019, 5:14 PM
>
> **Note:**
> "when you have spaces and gaps, the measurment is a smaller number because the marker isn't covering that part."

FIGURE 2.4   A star chart that characterizes learning of individual students

based on teacher use and feedback. Our data about teacher change (as discussed in Chapter 9) suggests that, over time, using these tools helps sharpen teachers' attention to developmental changes in key levels (and sublevels) of student thinking about measurement in the domains of length, angle, area, and volume. This shared knowledge eventually becomes a common language for describing learning that governs how teachers in the participating schools see and evaluate classroom evidence.

Studies of Student Learning

Both the design elements of the Measuring and Visualizing Space project and the iterative revisions that occurred through its duration were informed by learning studies, some taking place as the project unfolded, but others conducted before the project began by Lehrer and his colleagues (Knapp & Lehrer, 2005; Lehrer, 2003; Lehrer & Chazan, 1998; Lehrer & Pritchard, 2002; Lehrer & Slovin, 2014; Lehrer et al., 1998; Lehrer et al., 2003; Lehrer, Jacobson, Kemeny, & Strom, 1999; Lehrer, Jenkins, & Osana, 1998; Lehrer, Strom, & Confrey, 2002; Nitabach & Lehrer, 1996; Strom, Kemeny, Lehrer, & Forman, 2001). Like the present research, much of this earlier trove of research was in the form of classroom-based design studies, although they were smaller in scale than the project described here and more focused in scope. Most of these foundational studies investigated the development of student understanding of a single concept or small set of related concepts, such as how children interpret angle as bodily turn or how pacing can support students' understanding of iteration in the measurement of length. This body of existing research guided the initial design of curriculum units, which have now been iteratively and substantially revised to produce a coherent curriculum that builds cumulatively across the elementary grades. The revisions of existing activities and development of additional units have been informed by ongoing studies with the classrooms and students involved in the current research. As discussed in the previous section, one important source of evidence for guiding these revisions is the classroom-based and formative assessment data collected by teachers and entered into BASS via TOTs.

Exploratory Learning Studies

Throughout the project researchers conducted targeted learning studies to explore the potential of new instructional approaches that were being considered for incorporation into the curriculum design. In some of these studies, researchers initially co-taught with participating teachers during first iterations of the instruction or, alternatively, worked with small groups of students from participating classrooms. In others, the tasks were investigated in intact classrooms taught by participating teachers.

For example, Kobiela and Lehrer (2019) worked in a sixth-grade classroom, with Lehrer serving as the primary mathematics instructor, assisted by the regular classroom teacher. The emphasis of the instruction was on how dynamic production of area could ground students' conceptions of area and its measurement (Kobiela, Lehrer, & Pfaff, 2010; Lehrer & Slovin, 2014). Students explored area measure by sweeping a length through another length at an angle greater than zero—another instance of the project's strategy of positioning measure within the context of motion. In this case the students used squeegees to "sweep" vertical lengths through horizontal lengths to create quadrilateral areas of shaving cream spread on ceramic tiles. The resulting swept space was then dissected into units of measure, demarked with strands of uncooked spaghetti, by coordinating measures of the lengths of sides—in some cases, side lengths measured in different units, such as squeegee-inch. In this manner, area was presented as two lengths composed multiplicatively. The generation of area via motion was linked with the dissection and measure of the figures so created. Formative assessments during instruction and in interviews conducted two months afterward affirmed that students had developed dynamic images and meanings of area and its measure. Follow-up studies were conducted with the same materials in third- and fourth-grade classrooms at Mallard Elementary.

Following these studies with concrete materials, Brady and Lehrer (2020) developed an digital tool, implemented on iPads, to simulate work with the squeegees in a context that was both more flexible and considerably less messy. Students, in this case third-graders, began their work with the squeegees and tiles, but then moved to "sweeping" on screen. The virtual sweeping tool allowed students to more readily explore conjectures about changing units of length (including fractional units) and to observe the effects on a figure's area when those units are "swept" through each other. It also supported students in further extending their investigations of area by, for example, dissecting or partitioning a figure and rearranging the partitions to create alternative two-dimensional shapes with the same area as the original figure. Detailed videos of classroom instruction and student work products, supplemented by interviews with participating students, helped researchers understand the potential of these instructional approaches, including how to revise and expand them for general classroom use.

Yearly Interview Data

In addition to conducting studies like these, which provided detailed learning data on a narrow concept or set of related concepts, researchers sought to generate more systematic evidence about student learning across the domain of measurement more broadly, including information about how student understanding of measurement evolved across the elementary grades. To pursue this agenda, researchers conducted intensive, yearly, one-on-one, task-based interviews with a sample of participating students. These interviews were conducted during the first four years of the project at the Mallard site only, to establish the range of strategies that students were bringing to problems about measurement.

Interviewed students were nominated by teachers to represent a range of ability/performance levels in mathematics, with equal proportions of low-, middle-, and high-performing children. To the extent possible, students who participated in the initial year of data collection were also invited back for interviews in succeeding years. However, this longitudinal plan was only partially enacted, due to the highly mobile population. Many children selected for the first interview cohort eventually either left the school or were assigned in subsequent years to nonparticipating classrooms. Over the four years of the project, 236 students participated in these interviews; 46 of them were interviewed on two occasions and 24 on three occasions.

As Table 2.3 shows, during the first year of the project, researchers concentrated exclusively on students in grades one and two and focused on the measurement of length and (beginning in 2014) angle. As those students went on to the next grade, we added grade three in 2014 and 2015 and also extended downward to include kindergarten. Grades four and five did not come online until 2016, because the data collection plan required that the curriculum for those grades be based on the learning accomplishments and weaknesses established in the earlier grades. Therefore, curriculum design for the latter grades did not get under way until 2015, when the curriculum for kindergarten through grade three was reasonably well stabilized and analysis of learning data for those grades had been completed. Reflecting this development schedule, in 2016 interview tasks on area measurement were introduced in grade four and on volume measurement in grade five.

The interviews, which were administered in the spring of each academic year, lasted from 40 minutes (with kindergarten students) to approximately

TABLE 2.3 Number of Interviews Administered Each Year by Grade

	2013	2014	2015	2016	Totals
Kindergarten	0	22	20	20	62
Grade One	30	20	20	20	90
Grade Two	32	20	18	17	87
Grade Three	0	17	20	9*	46
Grade Four	0	0	0	19	19
Grade Five	0	0	0	22	22
Totals	62	79	78	107	326

* In 2016 the third-grade interview was focused exclusively on area.

90 minutes (for second grade and older) and assessed students' strategies on a wide variety of tasks intended to tap student thinking at each of the appropriate levels of the learning constructs. Some of these tasks were relatively brief (e.g., a student was asked to place fractional labels on a marked but unlabeled number line), whereas others were multi-step problems (e.g., a student was asked to write instructions to generate a square whose area was twice as large as that of a pictured square). Providing a sense of the scope of these interviews is the fact that the interview guide for the second grade (the longest interview) was 26 pages long. Major topics covered in the second-grade interview included *iterating and measuring angle as turn, properties of length units, measurement scale, partitioning, measure arithmetic, relational thinking, fractions and operations with fractions.* Each of these major topics included several subtopics. For example, the major topic *properties of length units* included assessment items intended to measure children's thinking about subtopics such as *identical units, tiling,* and the *inverse relation between unit length and measure.*

Formative Assessment Tasks and In Situ Observations

Further evidence of student learning was obtained as teachers collected data from all participating students on formative assessment tasks included in each of the curriculum units and intended to characterize student performance on the concepts taught in that unit. Depending on the number of units taught that year, the teachers collected, scored, and sent to BASS data from one to three semi-standardized formative assessments per year. The formative assessment tasks provided data on the performance of all participating individual students and, as well, provided teachers with a class performance profile. Supplementing the formative assessment data, teachers regularly converted their ongoing, everyday observations of children working on classroom activities into data by using the TOTs iPad system to connect observed student performances with appropriate levels of the learning constructs, as described in the section on assessment. This information, too, was collected via TOTs and sent to BASS for inclusion in new cognitive assessment models. Teachers met regularly in grade-level teams to review their class observation scores and discuss their criteria for scoring.

Studies of Teacher Learning and Practice

Studies of student learning were complemented by concurrent studies of teachers' learning. As we have described, the validity of the modeling conducted by the BEAR group at Berkeley depended in large part on teachers' uptake of the learning constructs as a shared framework. In this case, uptake included not only their willingness but also their capacity to both diagnose and respond to student thinking. Achieving this kind of shared vision required consistent commitment to understanding and working within the shared language of the measurement learning constructs. The project's professional development efforts were organized in large part toward (1) helping teachers achieve this vision and organize their professional conversation around it and (2) supporting them in using this shared framework productively to diagnose student thinking and respond to it appropriately in instruction.

Chapter 9 addresses teacher learning in greater detail. Here we will briefly explain the kinds of information that we have been analyzing to establish what and how teacher learning has made progress across the course of the project.

The first source of evidence about teacher learning and changes in teachers' instruction is the hundreds of hours of observational data that were collected both in professional development sessions and in classrooms. For instance, all the professional development sessions with teachers (typically, one two- to three-day visit per month and a week-long session during each summer) were carefully documented. These sessions included mathematics discussions led by researchers and visiting mathematicians, group analyses of student learning, and teacher questions about and feedback on learning constructs and curricula. In addition to observations of professional development sessions, there also are many hours of in-class observation of teaching in both individual teaching and embedded teaching formats. These visits are documented with video recording, supplemented with examples of student work. This documentary record provides evidence about changes over time in teachers' instructional practices.

More structured studies were conducted on a yearly basis to understand changes in teachers' understanding of student thinking. Each year, teachers participated in one-on-one interviews. During some of these interviews, teachers watched episodes of classroom teaching of measurement spanning grades one to five. The videos were selected from our archive of in-class observations. Teachers commented on what they noticed about core concepts of measure highlighted during the episode and what they noticed about student thinking and instructional practices. Teacher noticings were transcribed and analyzed to identify change over time in teachers' diagnostic skill and instructional repertoire for responding to commonly observed forms of student thinking.

All interviews included a series of questions in which the participant was asked to comment on issues about the project overall. They included (1) the "big ideas" in measurement that the teacher had addressed in the classroom during the past year, with retellings of episodes of classroom instruction that supported student learning about a big idea; (2) connections that the teacher had noted between measurement and other mathematical ideas; (3) how the teacher used formative assessment and ongoing observation of student learning in the classroom; (4) the teacher's interpretations of major forms of professional development, such as participation in embedded instruction and mathematical investigations, and their effects on the teacher's practice; (5) any collaborations on mathematics teaching that emerged either within the teacher's own grade or across grades; and (6) whether participating in this research/instruction project had led to any institutional changes, such as increased opportunity for professional collaboration, or to any tensions within the school or district, such as mismatches between standards and children's understanding of mathematics.

Having described the context for the design study and each of its essential elements, we turn next to describing how students understand the measurement of length, angle, area, and volume and connections of these measurement subdomains to rational number. We devote one chapter to each of these topics, articulating for each a theory of measure that details what and how students understand and how understanding typically evolves with appropriate instruction. Benchmark levels and exemplary performances are described in detail for each of the learning constructs. Examples of classroom activity clarify how instruction is designed to support students at these conceptual levels.

References

Bakker, A. (2019). *Design research in education: A practical guide for early career researchers.* New York: Routledge.

Brady, C., & Lehrer, R. (2020). Sweeping area across physical and virtual environments. *Digital Experiences in Mathematics Education, 7,* 66–98.

Carpenter, T. P., Fennema, E., & Franke, M. L. (1996). Cognitively guided instruction: A knowledge base for reform in primary mathematics instruction. *The Elementary School Journal, 97*(1), 3–20.

Carpenter, T. P., Fennema, E., Franke, M. L., Levi, L., & Empson, S. (2015). *Children's mathematics: Cognitively guided instruction,* 2nd revised edition. Portsmouth, NH: Heinemann.

Carpenter, T. P., Franke, M. L., Johnson, N. C., Turrow, A. C., & Wager, A. A. (2016). *Children's mathematics: Cognitively guided instruction in early childhood.* Portsmouth, NH: Heinemann.

Cobb, P., Confrey, J., diSessa, A., Lehrer, R., & Schauble, L. (2003). Design experiments in education research. *Educational Researcher, 32(1),* 9–13.

Davis, E. A., & Krajcik, J. S. (2005). Designing educative curriculum materials to promote teacher learning. *Educational Researcher, 34*(3), 3–14.

Kazemi, E., Gibbons, L., Lewis, R., Fox, A., Hintz, A., Kelley-Petersen, M. ... Balf, R. (2018). Math labs: Teachers, teacher educators, and school leaders learning together with and from their own students. *NCSM Journal of Mathematics Education Leadership, 19*(1), 23–36.

Knapp, N., & Lehrer, R. (2005, June). Changes in children's conception of spatial measure: Coordinating talk and inscription. Paper presented in Symposium *Understanding, building, and using symbolic representations of space and time* (M. Wiser, Organizer)). 35th Annual Meeting of the Jean Piaget Society, Vancouver, Canada.

Kobiela, M., & Lehrer, R. (2019). Supporting dynamic conceptions of area and its measure. *Mathematical Thinking and Learning, 21*(3), 178–206.

Kobiela, M., Lehrer, R., & Pfaff, E. (2010, April). *Students' developing conceptions of area via partitioning and sweeping.* Paper presented at the annual meeting of the American Educational Research Association, Denver, CO.

Lehrer, R. (2003). Developing understanding of measurement. In J. Kilpatrick, W. G. Martin, & D. E. Schifter (Eds.), *A research companion to principles and standards for school mathematics* (pp. 179–192). Reston, VA: National Council of Teachers of Mathematics.

Lehrer, R., & Chazan, D. (1998). *Designing learning environments for developing understanding of geometry and space.* Mahwah, NJ: Lawrence Erlbaum Associates.

Lehrer, R., Jacobson, C., Kemeny, V., & Strom, D. (1999). Building on children's intuitions to develop mathematical understanding of space. In E. Fennema, & T. A. Romberg (Eds.), *Mathematics classrooms that promote understanding* (pp. 63–87). Mahwah, NJ: Lawrence Erlbaum Associates.

Lehrer, R., Jacobson, C., Thoyre, G., Kemeny, V., Strom, D., Horvath, J. ... Koehler, M. (1998). Developing understanding of geometry and space in the primary grades. In R. Lehrer, & D. Chazan (Eds.), *Designing learning environments for developing understanding of geometry and space* (pp. 169–200). Mahwah, NJ: Lawrence Erlbaum Associates.

Lehrer, R., Jaslow, L., & Curtis, C. (2003). Developing understanding of measurement in the elementary grades. In D. H. Clements, & G. Bright (Eds.), *Learning and teaching measurement. 2003 yearbook* (pp. 100–121). Reston, VA: National Council of Teachers of Mathematics.

Lehrer, R., Jenkins, M., & Osana, H. (1998). Longitudinal study of children's reasoning about space and geometry. In R. Lehrer, & D. Chazan (Eds.), *Designing learning environments for developing understanding of geometry and space* (pp. 137–167). Mahwah, NJ: Lawrence Erlbaum Associates.

Lehrer, R., & Pritchard, C. (2002). Symbolizing space into being. In K. Gravemeijer, R. Lehrer, B. van Oers, & L. Verschaffel (Eds.), *Symbolization, modeling and tool use in mathematics education* (pp. 59–86). Dordrecht, Netherlands: Kluwer Academic Press.

Lehrer, R., Schauble, L., Strom, D., & Pligge, M. (2001). Similarity of form and substance: Modeling material kind. In D. Klahr, & S. Carver (Eds.), *Cognition and instruction: 25 years of progress* (pp. 39–74). Mahwah, NJ: Lawrence Erlbaum Associates.

Lehrer, R., & Slovin, H. (2014). *Developing essential understanding of geometry and measurement for teaching mathematics in grades 3–5.* Reston, VA: National Council of Teachers of Mathematics.

Lehrer, R., Strom, D., & Confrey, R. (2002). Grounding metaphors and inscriptional resonance: Children's emerging understanding of mathematical similarity. *Cognition and Instruction, 20,* 359–398.

Lewis, C., & Perry, R. (2017). Lesson study to scale up research-based knowledge: A randomized, controlled trial of fractions learning. *Journal for Research in Mathematics Education, 48*(3), 261–299.

Nitabach, E., & Lehrer, R. (1996). Developing spatial sense through area measurement. *Teaching Children Mathematics, 8*, 473–446.

Strom, D., Kemeny, V., Lehrer, R., & Forman, E. (2001). Visualizing the emergent structure of children's mathematical argument. *Cognitive Science, 25*, 733–773.

Wilson, M. (2005). *Constructing measures: An item response modeling approach*. Mahwah, NJ: Lawrence Erlbaum Associates.

Wilson, M. (2021). Rethinking measurement for accountable assessment. Keynote address, Australian Council for Educational Research Conference 2021, Sydney, Australia. https://doi.org/10.37517/978-1-74286-638-3_13

Wilson, M., & Sloane, K. (2000). From principles to practice: An embedded assessment system. *Applied Measurement in Education, 13*(2), 181–208.

3
ORIGINS OF QUANTITATIVE REASONING IN THE MEASURE OF LENGTH

By a "theory of measure" we do not mean a list of procedural rules for using tools to answer a measurement question. In contrast, we think of such a theory as a web of big ideas and procedures that, cumulatively, support the comparison of magnitudes of lengths, areas, volumes, or angles (Lehrer, 2003; Lehrer, Jaslow, & Curtis, 2003). This web is complex and multidimensional, and its components and their relations evolve as children grow older and experience appropriate instruction.

Note, too, that what we intend by a theory of measure is not a "theory theory" (e.g., Carey, 1985) with strong internal coherence among constituents, but instead an ensemble of concepts and procedures that emerges as students engage in comparisons of magnitudes (extents) of different attributes of space. A student who understands the role of unit tiling in generating a measure of the magnitude of a length may nonetheless need to revisit and work out implications of unit tiling for generating a measure of area. As students reconsider concepts like tiling for measures of different attributes, their theories exhibit increasing coordination among concepts within particular realms of spatial measure, as well as increasing coordination between realms, such as seeing different units of length and area measure as both providing means to tile a space. Hence theories of measure are similar to what a knowledge analytics framework refers to as coordination classes (diSessa & Sherin, 1998). These conceptual systems support increasing student participation in a practice of measure, meaning that individual contributions are increasingly coordinated with collective consensus about the appropriateness and properties of unit, origin and scale of measure, the values of reproduction of measure, and transparency about methods of measure, as well as a disposition to employ these ways of acting and knowing to measure

DOI: 10.4324/9781003287476-3

in new realms, such as those encountered in the study of natural systems (Ford, 2006; Rouse, 2015, Chapter 10, this volume).

Learning about measurement blends *practical activity*, such as how to use tools like rulers or protractors; the *conceptual underpinnings* of unit and scale, such as the role of a unit magnitude as a standard comparator for magnitudes (e.g., 10 m is 10 times as long as 1 m); and *symbolic systems*, such as the meaning of labeled hash marks in standard foot rulers to designate unit lengths and unlabeled hash marks to designate parts of units. These differing forms of knowledge-generating activity must be coordinated to develop a foundational understanding of measurement (Lehrer, 2003). Moreover, each form of activity is subject to collective consensus and critique, so measures of x are governed by shared understandings of the nature of x, what a measure of x means, and that the measure can be replicated by other measurers displaced in space and time (van Fraassen, 2008).

This and the next few chapters describe the big ideas and procedures that a theory of measure entails for each of four spatial domains: measure of length, area, volume, and angle. In addition, we provide similar information for the closely related domain of rational number when rational numbers are taken as measured quantities and as operations on measured quantities. At the most general level of analysis, we articulate children's thinking in these subdomains as traversing increasingly sophisticated levels of understanding that describe characteristic ways of knowing and doing. Most of these characteristic ways of knowing and doing do not arise spontaneously during the course of development but must instead be deliberately and systematically provoked and curated by the design of instruction. Some researchers refer to this kind of map as a learning trajectory or a developmental progression (e.g., Clements & Sarama, 2004, 2009; Simon, 1995; Simon & Tzur, 2004).

For example, in the learning trajectory for length that we will subsequently describe more completely, young children initially conceive of comparing magnitudes of lengths as a matter of aligning—directly comparing lengths to establish which is longer or shorter. Even direct comparisons are not transparent, because they involve negotiation about a common starting point for fair comparison. Shortly thereafter, children's invention of units enables them to construct quantities to describe the magnitude of a length. Quantities expand the conceptual scope of comparison to include additive (how much more) or multiplicative (how many times) relations between magnitudes of lengths. As students construct quantities, inevitably the need to fracture units to create more precise measures arises, and fractured units expand the scope of comparison to include non-whole-number quantities. Children's theory of length measure eventually encompasses relational thinking about measures and magnitudes. For example, if the magnitude of A is three times that of B, then the magnitude of B is necessarily one-third that of A. We regard benchmarks like these as generalizations about the qualities and sequence of typical conceptual development under conditions of effective instruction (Lehrer & Schauble, 2015). In our upcoming analyses, we refer to these midlevel generalizations as benchmark "levels" of thinking. Any one student may or may not strictly follow the sequence of understandings just outlined, as the levels are fluid and sensitive to different pathways of instruction. As with all models, this one is deliberately simplified; simplification supports instructional planning and assessment of progress at the classroom level of analysis.

The text in the next few chapters occasionally refers the reader to instructional guides described as units and lessons, which were developed to support the instruction in the research classrooms. We think of learning progressions or trajectories not as descriptions of invariant sequences of development, but as domain-specific pathways that emerge within a particular learning ecology. The units provide a vision of that learning ecology, although, of course, there are always some deviations between the planned and implemented curriculum (more on that issue in Chapter 9). We anticipate that these lessons may be useful to a broader community beyond our research participants; interested readers can find them at Routledge.com/9781032262734.

Overview of Children's Understanding of Length

We turn, then, to describing how young students tend to think about the measure of length and how their initial ideas serve as resources for further development. For each benchmark level that describes the development of children's conceptions of length, we outline a network of ways of thinking about unit and measurement scale that encapsulates that level, along with conceptual pivots that provide support. Conceptual pivots are instructional means that tend to instigate conceptual development, and, as we have explained earlier, in our program they lean heavily on metaphors of continuous motion to complement the more traditional focus on measure as discrete translation of units.

Although the blend of discrete translation and continuous motion as the foundation of quantitative thinking is revisited throughout each realm of measure (length, area, angle, volume), here we begin by overviewing some of the synergies afforded by these distinctive ways of thinking. Unit iteration makes evident that counts of units result in quantities and, more tacitly, that a measured quantity is a ratio of the magnitude (e.g., an extent) of an attribute to the magnitude of a unit. For example, a measure of 10 ft implies that the magnitude of the measured length is ten times the magnitude of 1 ft. Unit iteration also invites thinking of composite units as a way to generate more efficient counts. For example, translating a 12-inch ruler to measure a 26-inch length enables a more efficient count (2 ft 2 in, or 26 in) than counting on from 0 at each translation (e.g., 0–12 in, 0–12 in, 2 in). The motion metaphor complements unit iteration by treating an attribute as generated by continuous movement in one or more dimensions. For example, a magnitude of length is a path generated by continuous movement in one dimension from start to finish (Abelson & diSessa, 1981). The conceptual leverage of this metaphor is that it helps children interpret 0 as the starting point of the movement, a reference point that must be symbolized and shared, so that 0 indicates no distance traveled, rather than the initial oddity of a ratio of iterated length to unit length of 0. (Children often imagine that one would be a better expression of the origin, because it denotes the first unit to be counted in the measure of the length.) Other conceptual bootstraps of the length-as-path metaphor include helping children understand whole-number units and fractional units as locations along a path. This metaphor also helps children interpret values of fractions greater than one, such as $\frac{7}{2}$ ft, as distances traveled, and in this instance as equivalent to $3\frac{1}{2}$ ft, when students coordinate the same magnitude of length in two distinctive units of measure, ft and $\frac{1}{2}$ ft.

Benchmarks in Thinking About Length

Given this overview, we turn now to articulating in greater detail a trajectory of emerging conceptions of length measure and conceptual pivots (forms of instruction) that tend to support the development of these conceptions. The trajectory is structured as six distinctive ways of knowing (articulated as benchmarks or levels) and of related supports for learning (conceptual pivots), with an overall aim of portraying an ecology of knowing. We begin with the level that describes how children tend to think about length as they are just beginning formal instruction in measurement.

Directly Comparing Magnitudes

Measurement compares magnitudes (extents) of an attribute; when we look at a length, we are perceiving its extent or magnitude. To initiate measurement as the comparison of perceived magnitudes, children are challenged to compare magnitudes of lengths directly. As the name implies, direct comparison juxtaposes or aligns magnitudes to facilitate comparative judgments about more or less, but without yet quantifying how much more or less (Piaget, Inhelder, & Szeminska, 1960). For example, standing back to back can reveal which person in a pair is taller. As we will explain, representations of magnitudes can be constructed too, so that direct comparisons can encompass representations of lengths (Hiebert, 1981). For example, paper strips may be used to represent the heights of children, and the strips can be directly compared.

Although direct comparisons seem straightforward and transparent, they are but surface manifestations of a web of relationships that undergird much of measure. For a start, measures should answer a question of comparison. Sensible and easily understood questions motivate a youngster's first encounters with measure, so that comparison is understood as the guiding rationale. For example, in a kindergarten class, children looking at a small collection of pumpkins wondered, which one is biggest? But as children considered this question, they noticed and mentioned multiple senses of *big*. Some children were thinking about tall pumpkins, others about pumpkins with the greatest girth (fatness), and yet others about pumpkins with the most heft (weight) as biggest. To make sense of the question, children had to decide which features of the pumpkins could help them decide about "biggest." Which attribute or attributes would be the basis of comparison? Contextualizing measurement in this way helps students grasp that the attribute one measures depends on the question or purpose being pursued. To emphasize that measurement should be guided by one's questions, it is worthwhile, at least on some occasions, to encourage children to propose and evaluate attributes that could be measured to best answer the question(s) that the class is entertaining.

After children decided that "bigness" would mean comparing pumpkin heights and girths, another challenge arose. Heights could be checked by physically manipulating the pumpkins directly—for instance, by placing three pumpkins side by side and visually noting which was tallest. But what about girth? The teacher asked if string or strips of adding machine tape could be used, and several children suggested wrapping the tape around the pumpkin's midsection (a proposal that led to further discussion about how to identify the midsection). Using paper strips to stand in for a length shifted the grounds of appraisal from physical to representational comparison, although the comparison remained direct in that children could literally juxtapose the strips to compare heights or girths. Using intermediate representations was more challenging than simply comparing the pumpkins directly because it required children to hold in mind the relation between the objects of interest (the pumpkins) and the representations (the paper strips) of their heights and girths (circumferences). This mapping between referent and representation is not always straightforward, especially when the representations do not physically resemble the focal attribute in all respects. For example, some of the kindergartners were surprised to observe that a paper strip wrapped around the girth of a pumpkin could be unwrapped to form a straight length, just as in the strips representing height.

Having decided to use tape to represent height and girth, children worked in small groups to reproduce the height and girth of each pumpkin with paper tape. This meant that they could now compare heights and girths without having to move the pumpkins because they could simply compare the strips. But further difficulties emerged when their teacher displayed the strips that different pairs of children had cut to represent the height of the same pumpkin. It was clear that some of the strips were longer than others, even though they were all purportedly representations of the same pumpkin's height. Why did this happen?

The teacher next asked children to describe *how* they had gone about the process of measuring height, and as each group described their process, the teacher reenacted it so that everyone could observe the measurement process in action. One group explained that they cut a strip to represent the height from the table to the top of the pumpkin stems. Another reported that they measured from the table to the highest point on the pumpkin bodies. This group decided not to include the stems because one pumpkin did not have a stem attached. Rather than "measuring straight" from the table to the tallest point, a third group bent their strips to mold them around the curve of the pumpkins. As children explained and critiqued each other's strategies, they eventually concluded that they could not achieve a "fair comparison" without agreeing on a standard way to measure attributes like "tallness" and "fatness." Attributes require a common definition so that magnitudes can be compared. A more subtle aspect of defining attributes is that they must be measured in the same way. For example, even after agreeing that girth should be measured at the fattest part of the pumpkin, some children stretched the tape tightly while others used a more relaxed grip. As a result, the lengths of tape differed in spite of children's agreement about where and what to measure. These differences in processes of measure had to be resolved, and as they debated how to do so, children were involved in coming to see the virtues of reproduction in measure. Reproduction in measure is a collective value that is made feasible by adopting common definitions of attributes and processes of measure.

Once the attributes were defined and a common way of representing each was adopted, the stage was set to compare the heights of the pumpkins. The teacher motivated this comparison by randomly affixing strips to the board in an unstructured display. She then asked children, "How should we arrange them to decide which is tallest?" As she had intended, one of the children responded that it would be difficult to compare the heights unless the "tops of the strips" were all aligned at a common level. This concept of baseline (or zero point when a quantitative measurement scale is used) is another fundamental principle of measure that may or may not occur to young children. A common origin was tacit when the teacher first held the strips representing the height of a single pumpkin, but she now deliberately problematized this idea to enhance the likelihood that students would grasp the underlying logic of adopting a common baseline for comparing lengths.

The teacher also knew that symbolizing seriation would not necessarily be obvious to all children. How do we say and write (symbolize) ordered relations? The teacher introduced a notation that would communicate the relations that children were expressing in words. For example, "Pumpkin C's height was greater than pumpkin D's height" was notated as C gt D (where gt represented *greater than*). As there were different orderings for height and girth, this conversation also helped students recognize that the order of magnitudes for the pumpkins depended on the attribute selected.

In sum, although children were ostensibly engaged in the very simple enterprise of directly comparing perceived magnitudes of attributes of pumpkins that everyone could easily see, the class debated and resolved disagreements about what to measure, how to measure, how to compare representations of length (emphasizing the idea of baseline/zero), how to order heights and girths based on relative lengths of these representations, and how to symbolize the results. These fundamental principles of measurement were raised in the context of children's activity organized around a simple question of comparison. The teacher felt they were worth exploring because she knew that those same principles would emerge again repeatedly as children encountered new contexts for measuring and as they further elaborated the theory and practice of measure. Hence, learning about measure was supported by a dialectic between individual enterprise and collective endeavor.

Table 3.1 an abbreviated version of Table 2.2, illustrates the indicators of the conceptual underpinnings of *direct comparison of lengths* that students formulate when they receive the kind of supportive instruction exemplified in this classroom example. We employ the term *indicators* to suggest that children construct mental representations, initially of local scope, that guide the performances described in the second column of the table. Performances like these are visible in classrooms and are informative about conceptual progress.

TABLE 3.1 Indicators of Conceptual Underpinnings of Direct Comparison

Description	*Examples*
Distinguish (e.g., equal, not equal) or order (e.g., greater, lesser) magnitudes of an attribute by direct comparison of representations. Symbolize comparisons.	"Pumpkin A is taller than pumpkin C" (Student aligns paper strips that stand in for height with common origin and notices that A's strip is longer than C's strip). $H_A > H_C$ (Student interprets and chooses appropriate pair of strips representing this relation.)
Distinguish (e.g., equal, not equal) or order (e.g., greater, lesser) magnitudes of an attribute by direct physical comparison.	"This book is taller than that one" (Student aligns the books and compares).
Define the attribute being measured so that measures can be compared.	"Fat means how far it is around the caterpillar" (student makes analogy to circumference of her wrist).
Identify measurable attributes (qualities).	"We could find out how long the caterpillar is or how fat it is."
Pose a question or make statements about a potentially measurable object of interest.	"Which caterpillar is the fattest?" "Which pumpkin is tallest?"

Table 3.1 is organized at a midlevel of description that lies between very fine-grained, microgenetic moments of a child's activity and an overarching, more general and abstract description of an entire benchmark level, here "directly comparing." The indicators in Table 3.1 are organized to represent later developing conceptions in the upper region and earlier developing conceptions in the lower region. The ensemble of conceptions collectively constitutes the benchmark.

Explaining How Properties of Units Affect Measure

Direct comparison enables children to order magnitudes of length but does not support comparison of *how much* more or less. To support these comparisons, quantities must be constructed, and to do so, children must learn about properties of units and how to accumulate units to produce a measured quantity. Magnitudes of length are elaborated as counts of unit and, with common units, are easily compared (Piaget et al., 1960; Thompson, Carlson, Byerley, & Hatfield, 2014). These conceptions about the properties of units characterize the second benchmark level of the learning trajectory for length measure.

There are several simple instructional ways to initiate children's encounters with properties of unit. These encounters will come to have wide application, as ideas about unit are fundamental to all domains of measure (see unit 2, Length). One of these instructional tasks is to have children compare the lengths of two paths that cannot be directly aligned but that are engineered to have a whole-number measure, such as 8 marker lengths or some other ready-at-hand unit. As with instruction at the previous direct comparison level, a question of comparison should motivate activity. Examples are "Which is longer?" and then "How much longer?" Children can decide how to use markers, pencils, or some other unit to find out. Questions of *how much* can be answered by counts of units. During the course of activity, opportunities will arise to consider that units should tile a length so that gaps and overlaps between units are eliminated. Some children may introduce gaps as they lay markers down, or if they don't, teachers can follow up by introducing gaps and asking children to predict what will happen to the measure and why. Children should not just rotely code these ideas as unexamined "rules" of measure ("no gaps, no laps"), but should instead come to understand that gaps between units produce underestimates of measure, because some of the space is not accounted for, and overlaps produce overestimates of measure, because some of the space is accounted for more than once. In a task like this, units are easily counted to produce a measure, and teachers encourage children to notate the measure with numeral and unit, as in 8 m means 8 markers long.

Measures always have two aspects, one rooted in practical activity and the other in symbolic structure. Here both are simple, but they do lead to new challenges: How can the measures be compared? How can that comparison be expressed? Responding to the first question helps children understand that the quantities they have constructed correspond to visible differences in length, such as 8 m vs. 6 m. This is an important principle of measure: Measures correspond to states of a property, so if one length's magnitude is greater than another's, its measure should be greater if the unit of measure of both is the same. A second outgrowth of quantitative comparison is its relation to arithmetic expression. For example, first-grade children expressed the arithmetic relation between path lengths of 6 markers and 4 markers as $4 \text{ m} + 2 \text{ m} = 6 \text{ m}$ (thinking about counting on from the measure of the shorter path to the measure of the longer path), and as $6 \text{ m} - 4 \text{ m} = 2 \text{ m}$ (the difference in length measures), and also as $6 \text{ m} - 2 \text{ m} = 4 \text{ m}$ (counting down from the measure of the longer length to the measure of the shorter length). Chapter 7 in this volume further addresses how measurement can serve as a context for linking to and conceiving of the meanings of arithmetic operations.

After the first graders counted marker units to construct the measure of a path made with painter's tape on the floor, the teacher gestured and asked, "Where's 1?" Children laughed and pointed to the first marker that was tiled on the path. The teacher then invoked the metaphor of travel, using a toy bus with its front placed at the start of the length. She started "driving" the bus along the length, asking repeatedly, "Have I traveled 1 marker yet?" Stopping at the midpoint of the first marker, she asked, "Here? Has the front of the bus reached the end of the first marker?" Eventually, the children decided that counts of 1 m, 2 m, etc., were indicated when the front of the bus had reached the endpoint of each marker. Here motion helped children understand the unit as signifying a distance traveled, so that the measure represented total distance, not simply a count of a collection of units. This is an important distinction between collection metaphors of number (e.g., count of objects in a container) and measure (count as signal of distance traveled), and it is a distinction children need assistance to grasp (Lehrer et al., 1998; McClain, Cobb, Gravemeijer, & Estes, 1999).

As instruction continued, the teacher asked which numeral should mark the location of the start of the bus travel. Some suggested the numeral 1, but that had already been reserved for the endpoint of the first unit. Others suggested 0, because no distance had yet been traveled and the bus was stationary at that point. This made sense and was adopted by the class as a new convention. As mentioned previously, children often have difficulty understanding 0 when measure is approached purely as a matter of ratio (e.g., 10 u is ten times as long as 1 u, and 0 u is 0 times as long as 1 u).

Other instructional support helps children develop related conceptions of properties of units that support conceiving of length as a measured quantity.

For example, children use a collection of units of different lengths (multiple copies of each unit, such as longer and shorter paper clips) to measure a length such as an edge of a table or desk. This situation can reveal children's conceptions of tiling and how they treat mixtures of units of different lengths. Mixtures of units are acceptable if the units are labeled, because other people can then reproduce the measure. It is helpful for kindergarten and first-grade students to attempt to re-create a length given its measure (e.g., "My line is four large and two small paper clips long. Here is a bag of paper clips of two different sizes. Can you copy my line in your notebook?") The emphasis on reproduction without access to another measurer's activity is a way to help children develop an appreciation of the role of communication and the importance of the reproduction of a measure. Often children presume that their activity is transparent to others displaced in space and time, and therefore they fail to label different units (e.g., they may record four large and two small paper clips simply as 6, or six paper clips).

The measure of a magnitude is a ratio of the magnitude of the measured length to the magnitude of the unit length. As a result, the count of units is affected by the magnitude of the unit length. When compared with longer unit lengths, shorter unit lengths result in a greater measure of the same magnitude, and when compared to shorter units, longer units result in a diminished measure (Carpenter & Lewis, 1976; Lehrer, Jenkins, & Osana, 1998; Sophian, 2007). Some researchers feel that this inverse or compensation relationship is challenging for students to understand even through the late elementary grades (Hiebert, 1984; Szilagyi, Clements, & Sarama, 2013), but it can be made accessible and visible to young children. For example, children can experience this relation by measuring the distance between two landmarks with different units. Comparing heel-to-toe walks with larger and smaller feet evokes questions like: Why does the count of small feet exceed the count of large feet for the same distance traveled?" Thompson et al. (2014) suggest that an explanation requires holding in mind the distinction between a measured quantity and its segmentation into unit-parts, so that different measures of the same perceived extent of length are sensible in light of use of different units.

Measuring distances with standard units, like the length of a teacher's foot, helps children understand how standard units facilitate comparisons of different lengths. If children are invited to choose the most appropriate tool to measure short and long lengths—such as their desk and then an outdoor path—they begin to develop a sense of the suitability of a particular choice of unit for the distance being measured. Thus, yards are a better unit than inches for measuring the length of a path in a field or the distance in a hallway. Reasoning about matters like these helps students appreciate how properties of units affect the generation of measures. This appreciation, in turn, supports more refined comparisons of magnitudes of length, to answer questions such as how much more or less. Table 3.2 illustrates some of the performances associated with these conceptual underpinnings of unit, again arranged so that upper regions indicate concepts that typically build upon and emerge later during the course of learning. Understanding the roles and implications of unit for measure is again constituted by the network of relations among these constituents.

There is some dispute in the literature about the value of allowing students to work with nonstandard units, such as paper clips or markers, before moving on to tasks that feature the use of standard units and measuring tools. Some (e.g., Boulton-Lewis, Wilss, & Mutch, 1996; Clements, 1999) have pointed out that children can often measure correctly with a ruler before they are able to deploy a measurement approach that involves nonstandard units. They further argue that using standard units is less demanding (because students do not need to keep track of the multiplicative relationships of different units to the length being measured) and is more meaningful from a real-world perspective (Clements & Battista, 2001). The disagreement appears to be over two different ideal sequences for instruction that are associated with differing perspectives on the role of measure in early mathematics. In the first sequence, children initially learn to measure with standard manipulative units or rulers

TABLE 3.2 Indicators of Conceptual Underpinnings of Unit

Description	Examples
Qualitatively predict the inverse relation between size of unit and measure.	"If we use small steps, the measure (of the same length) is more than if we use large steps." "It will take fewer of Ms. H's feet to travel from here to there (points) than it did for my feet."
Consider suitability of unit and explain why.	"That (distance) is very long, so using my clipboard (as a unit) works better than using my pencil (as a unit)."
If units are not identical, distinguish among them by labeling units.	Tiles 2 blue, 4 red units. Reports measure as "2 blues, 4 reds." Tiles length with larger and smaller paper clips. Reports measure as "4 big paper clips and 3 little paper clips."
Use identical units and explain why.	"It is better to measure with all the same units because then you can just count the number of units to find the measure."
Tile units and explain why.	"The units should touch so there is no gap. Gaps mean that some of the space is not measured." "If they (units) overlap, you measure the same space twice."
Tile units (no gaps, laps).	Student aligns a collection of units along a length without gaps or overlaps. Measure is the count of the collection.
Associate measure with count.	"This book is 4" (student reads number off a ruler). "The pencil is 5 paper clips long" (the paper clips may not be identical lengths).

and, once those early skills have been mastered, are then considered ready to address the problem of determining how a measure changes when the unit lengths change. In contrast, in the approach advocated here, children devote more consideration from the beginning of instruction to grappling with what it means to measure. As we have explained, early experiences are motivated by questions of comparison, which require children to address questions like "How much longer?" Questions like these motivate partitioning, not only into units, but also into part-units via splitting or equipartitioning, to increase precision while addressing comparative questions (e.g., Confrey, Maloney, Nguyen, & Rupp, 2014). This early emphasis on partitioning is consistent with our emphasis on using concepts in measure to provide a foundation for children's evolving understanding of rational number (e.g., Confrey, Maloney, Nguyen, Mojica, & Myers, 2009), a topic that we examine in greater detail in Chapter 7. Consistent with this goal, students experience an early emphasis on formulating and symbolizing relations between a length and the unit used to measure it (e.g., a measure of 10 u means that the magnitude of the length is ten times the magnitude of the unit length, and that with choice of a different unit, the measure will change). Rulers appear in instruction as tools that incorporate solutions to the problems of measure once children have already articulated them—that is, problems associated with the development of tiling, equipartition, zero point, and scale. Tools like rulers are framed as providing ways of solving dilemmas that children acknowledge as problematic because they have previously met and tackled them (Lehrer, Jacobson, Kemeny, & Strom, 1999; McClain, Cobb, Gravemeijer, & Estes, 1999). Thus, children are in good position to understand both the meaning and the value of the solutions that conventional tools provide. Ultimately, the goal is for students to understand and appreciate the decisions that guide conventions, not just to be able to use those conventions fluently.

Unit Iteration and Constructing a Measurement Scale

As students come to understand properties of unit that afford measure, an important transition is to conceptualize a unit as a length that can be translated, so that a measure of the magnitude of a length can be established by unit-length translation. Translation connects measurement to the more general study of isometries—transformations that preserve the lengths and angles of a figure. Translation of units is called iteration, and it is signaled in practical activity by reuse of a unit to measure a length (Clements, 1999; Clements & Stephan, 2004; Lehrer, Jaslow, & Curtis, 2003). Fluent unit iteration signifies that the student can imagine the partitions of a continuous length and can mentally represent them even after the unit has been moved and is therefore no longer visible to demark those partitions (Piaget et al., 1960). Moreover, unit iteration embarks students on multiplicative comparison—for instance,

one length's magnitude is four times that of a unit length, which can be practically established by noticing the congruence of 4 copies of the unit length
and the length with a measure of 4 unit lengths.

One way to help children conceive of units as iterable is to pose the challenge of measuring paths, but now with only one unit length, such as a wooden
dowel, available. Children typically assert that doing so is impossible and will
often protest "There aren't enough units." In such cases, providing two units
to get them started allows children to experiment with end-to-end placement
of the units so that one unit length remains fixed and the other is translated.
This procedure is a stepping stone to reusing a single unit while attending to
ways of marking the endpoint of the unit. Like other measurement concepts,
iteration is constituted both by pragmatic activity—translating a unit in space
and marking endpoints as starting points for the next iteration—and by mental
generation of a unit length.

To anchor iteration to the properties of unit emphasized in the previous
level, we observed a second-grade teacher challenge students to use a standard
unit, a cardboard version of her footprint, to construct a 4 TF (teacher-foot)
path on the floor. As students worked, the teacher video-recorded how different pairs of students iterated the foot unit. Each pair marked the beginning
and end of their path length and cut a paper strip in a length congruent with
the path length. Students expected that all the lengths would be congruent
(within a smidgeon due to imprecision of the paper cutting). When the teacher
collected all the paper strips and held them up to view, most were very nearly
congruent, but some were considerably longer or shorter. As they reviewed
the video records, students revisited their procedures for marking endpoints of
unit intervals. For instance, using a finger to mark the endpoint of a unit created unanticipated gaps between units, which led to lengths longer than 4 TF.
Diagnosing this problem inspired children to develop alternative strategies for
indicating endpoints of unit intervals and again emphasized measure's blend of
ideas and practical activity.

Building on experiences like these with unit iteration (see Length unit 3),
teachers also engaged students in the design of tape measures that they
then used to measure different lengths. Designing tape measures involves constructing a scale of measure, such as the one exemplified by rulers. For children
constructors, design challenges include the following: Where should numerals be placed? What do they mean? Why should the tool employ a standard
unit (e.g., a teacher-foot)? How does one use a scale to measure? Here again,
thinking of length as distance traveled helps make intelligible the placement
of numerals on a measurement scale. Teachers encouraged children to imagine
movement by sliding their fingers along the tape measure and pausing at the
endpoint of each unit. As it did at the previous level, movement consolidates
the idea of length as distance traveled so that it makes sense to start a measure at 0 and to place numerals on the tape measure to indicate endpoints of

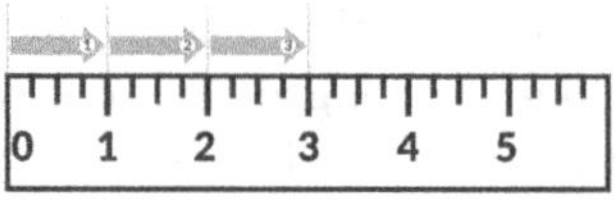

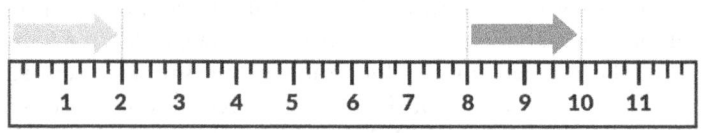

FIGURE 3.1 Scales structure units as distance traveled

unit intervals. Concepts of a measure as accomplished by unit iteration and as indicating a location along a path are coordinated to constitute a measurement scale—a way of specifying relations among units to mark quantities, as shown in Figure 3.1.

An important extension of unit iteration is to expand translation to composite units, such as 2 iterations of 5 units to construct a length with a measure of 10 units. Evidence about this kind of thinking is indicated by how a child uses a tape measure to obtain a measure longer than the tape. A child who is thinking about composite iteration will translate the tape measure and count on from the accumulated measure of the tape to find the measure of the magnitude of the extended length. For example, to measure a 7-unit length with a 4-unit tape measure, the child starts the count at 4 and then translates the tape and continues to count, 5, 6, 7, thus treating 4 as the zero point (the origin) of the second iteration. As children iterate composite units, the idea that any point can serve as the origin or the zero point arises. (This recognition is extended at the next benchmark level to include any point on a scale as a suitable starting point, as shown in Figure 3.2.)

Quantities continue to afford comparison of lengths, with how much more or less expressed by arithmetic operations, such as 10 u − 5 u = 5 u. Augmenting lengths by joining or gluing them together can be expressed by addition, as in 10 u + 5 u = 15 u. The metaphor of addition as joining has a motion counterpart: traveling 10 u on a path and then continuing to travel another 5 u, ending at 15 u. Multiplicative comparisons of quantities are inherent in any measure. For instance, a measure of 10 u is ten times as long as 1 u. This multiplicative comparison can be validated with unit iteration: 10 iterations of 1 u construct a length of 10 u. Multiplicative comparisons are supported by using specific forms of talk in classrooms (Thompson et al., 2014), especially the phrase "times as long as." The interplay between multiplication and iteration is supported by a language of "copy," as in 10 copies of 1 u make a length with measure 10 u, so a length with measure 10 u is ten times as long as

![ruler diagram with scale 1 through 11 and two gray arrows]

FIGURE 3.2 The origin of measure is arbitrary

the unit length. Copying can be enacted and anchored to symbolic expression, so that 10 u is re-expressed as 10×1 u. Moving across these symbolic forms helps children coordinate talk, activity, and notation to develop more robust conceptions of multiplicative comparison. To reason in this manner, children must again see, as they experienced literally in their previous investigations, that a measure of 10 u implies a subdivision of a length into 10 unit lengths. A second, more challenging form of multiplicative comparison involves composite unit lengths, as in 10 u is two times as long as 5 u, so 2 iterations of 5 u construct a length with a measure of 10 u. Table 3.3 describes visible performances associated with students' emerging conceptions of unit iteration and unit-scale construction.

2-Splitting and Symbolizing 2-Split Units as Measures

As the markings on a traditional ruler illustrate, fracturing units allows one to find the measure of lengths that are not a whole number of units long. Splitting (Confrey & Smith, 1995) arises from everyday actions of breaking or fracturing lengths of objects, such as sticks or candy bars, but splits of a unit have the additional quality that the resulting parts are congruent—they are equipartitions. Equipartitioning is foundational to thinking about fractions (Confrey, Maloney, Nguyen, & Rupp, 2014). We introduce children to splitting via fracturing by factors of two, beginning with a 2-split of 1 unit. The need to split a unit arises as children use tape measures to measure lengths that are staged to have whole-number measures and half-unit measure with standard units, such as the length of a teacher's foot. Splitting by 2 and using the language "half of" appear to be within the realm of children's everyday thinking. To extend these everyday conceptions to measure, 2-split part-units are iterated to measure lengths (Lehrer et al., 1999; Lehrer & Pfaff, 2011). For example, a magnitude of length might have a measure of $\frac{5}{2}$ u, and the measure is found by iterating $\frac{1}{2}$ u five times. This means that children are now in a position to interpret $\frac{5}{2}$ u as 5 copies, or iterations, of $\frac{1}{2}$ u. An additional challenge is to coordinate measure of the same magnitude of length in two different units, so that $\frac{5}{2}$ u is the same distance from zero as is $2\frac{1}{2}$ u. When 2-splits are composed, as in one-half of one-half, the resulting number of parts extends the possibilities of part-unit iteration to fourths and eighths. Composition of 2-splits constitutes a gentle introduction to exponential functions that students usually do not encounter for years, but that do seem well within their grasp. For example, students can reason that the number of parts increases (2, 4, 8, 16, …) as 2^n where n represents the number of splits. As splitting units into parts comes into focus, measure provides contexts and ways of thinking that are surprisingly powerful for bootstrapping children's emerging ideas about rational number. These connections are elaborated in greater detail in Chapter 7 of this volume.

TABLE 3.3 Indicators of Conceptual Underpinnings of Unit Iteration and Scale

Description	Examples
Iterates composite unit (e.g., a 3 in length is thought of as 3 1 in lengths and as a length with a measure of 3 in).	"I lined up the zero on my ruler with the start of the path and traveled to the 3 in mark at the end of the ruler. Then I kept my place and moved the ruler to line up my place with zero, and the end of the path was at 2 in. So, the total distance that I traveled was 5 in." (Note: simultaneous mental marking of a location as 3 in and as 0—the start of the next unit.) 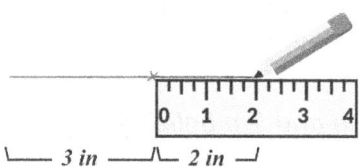
Explain how a measure is the ratio of the number of units accumulated to the unit length.	"This is 5 Goades long. It is 5 times as long as 1 Goade." "Here's a length. Draw a length that is 4 times as long." (To solve, student iterates the length 4 times to draw a line 4 times as long.)
Use and justify standard unit (including conventional unit) as facilitating comparison of lengths.	"If we all agree to use Justin's foot as a unit, it's easier to compare our measurements of the lengths of different desks." "When we order online, we all need to use the same units, like feet and inches, so that everyone knows what we mean."
Symbolize/write units at endpoints of unit intervals on measuring tool (ruler, tape measure) to indicate distance traveled from origin.	"You don't write a 1 in the middle of the unit like this one …  … because the unit starts at zero and ends at 1—that's how far you have traveled! When you put it in the middle, you don't see where the unit ends." 
Symbolize/write starting point of measure as zero (0).	"Here (points to starting point on tape measure) is zero." Labels as "0" or says "zero." *Note: The student's placement of the symbol, 0, may not align exactly with the origin due to materials used.*
Re-use (iterate) a unit to measure.	"I just had one unit so I marked its end and then used it again, marked its end again, and kept doing that. It's 8 paper clips long." *Note: Iteration includes both the concepts of translation and accumulating count, as well as the procedural competence involved in keeping track of the translated unit.*

We turn now to some of the conceptual pivots—forms of instruction that tend to promote part-unit iteration (see Length units 3–7). As described previously, teachers usually stage the need to split units. Naturalistically, many different splits are possible to measure a non-whole-number unit length, so teachers engineer contexts where one-half of a standard unit (e.g., a teacher-foot length or a conventional foot) is responsive to answering the question "How much more?" for the segment of length that extends beyond whole-number measure of the length. The standard unit is rendered as a paper strip so that it is easy to fold. This situation prompts consideration of how to construct *half of*, which is easily done by folding the paper so that the ends match. The teacher enacts half-unit iteration with a single $\frac{1}{2}$ unit to introduce an additional property of half-unit: 2 iterations of the half unit construct a unit length. Other 2-fractures of the unit do not—and children try different partitions (e.g., a unit partitioned into 2 non-congruent parts vs. a 2-split part) to verify this relation between iteration and the definition of $\frac{1}{2}$ unit. This experience provides children with a practical tool to evaluate a claim that a part of a unit constitutes one-half of the unit, even if they have not personally constructed it by 2-splitting a unit length. For example, a teacher holds up a unit length and children cut a strip of paper to estimate one-half of that length. Two iterations of the cut length should be exactly as long as the unit length. This iterative principle can be extended to consider any prospective split of the unit (Cortina, Višňovská, & Zúñiga, 2014).

Once students have a firmer grasp on the meaning of one-half unit, teachers instigate part-unit iteration by having children unfold the unit length, grip the paper between thumb and forefinger, close their eyes (so that they need to rely on touch, which highlights continuity), and travel with their fingers from 0 to one-half unit. Then, they start again from 0 and travel to two-halves unit, and then gesture with eyes closed at a distance three-halves of the unit length. As they travel, students repurpose iterative counting with whole numbers of units to count by half-units. Then students iterate a half-unit to measure the path that first raised the need to split the unit, counting, for instance, nine half-units or four and one-half units. Imagining a length as a distance traveled, a concept that was consolidated earlier in instruction, helps students make sense of fractional quantities greater than one as measured quantities.

Continuing with instructional support for half-unit iteration, students are introduced to conventional notation for fractions, $\frac{a}{2}$, where the denominator represents the number of equal parts of a unit—its *splitting number*—and the numerator, *a,* expresses the *number of copies* of one of the 2-split parts. With these emerging understandings, students go on to revise their whole-number tape measures to include locations for $\frac{1}{2}$ unit. The metaphor of travel again helps students understand why $\frac{1}{2}$ units are marked at the endpoint of the half-unit interval. Students use their reconstructed tape measures to measure paths with half-unit measures, some of which are staged to include magnitudes of path lengths greater than the magnitude of the length of the tape measure. For example, two first-grade children used a 5-unit-long tape measure, with $\frac{1}{2}$-unit intervals marked, to measure a path that was $5\frac{1}{2}$ units (usually this is a focus of second-grade instruction, but these ideas are within the grasp of many first-grade children if they have had previous experience with investigating properties of units and whole-unit iteration). To do so, they first measured the length as $\frac{10}{2}$ units and marked the endpoint of the tape measure (at 5 units) and then translated the tape another $\frac{1}{2}$ unit to obtain a measure of the length as $\frac{11}{2}$ units. Note that their strategy indicated a firm grasp of conceptual foci originally introduced in the previous level, especially composite-unit iteration. Another student volunteered that the measure could also be called $5\frac{1}{2}$ units. One of the challenges of coordinating half-unit and mixed-number (whole-unit) representations is that it requires a child to hold two levels of unit in mind simultaneously, both the unit-split (that is, the $\frac{1}{2}$ u) and the original unit length. It helps if both levels are in view at the same time on the tape measure.

Instructional support for unit-partitioning (as in Length units 4–7) can be extended to compositions of 2-splits to produce fourth-units by splitting a unit length by 2 and then by 2 again. If the splitting is performed with paper-strip units, then children can readily see that the result is 4 congruent parts of one unit. And, if it is 2-split again, then 8 congruent parts are produced. Acts of paper folding are complemented by their symbolization as scalar multiplication (e.g., $\frac{1}{2} \times 4$ u, $\frac{1}{2} \times \frac{1}{2}$ u), where fractions have an additional role as transforming or operating on measured quantities. That is, constructing a part-unit is the result of a mathematical operation, so there is a way to coordinate hands and mind.

Part-unit iteration again establishes that $\frac{1}{4}$ u is the only part-unit for which 4 copies reconstruct the unit length, and subsequently children check predictions about the number of copies of $\frac{1}{8}$ u necessary to reconstruct 1 u. Other fractures of 4 or 8 do not have this property. Students again revise their tape measures to indicate locations of fourths and eighths. With travel of distance firmly in mind, students iterate by $\frac{1}{4}$ u, counting $\frac{1}{4}$ u, $\frac{2}{4}$ u, $\frac{3}{4}$ u, $\frac{4}{4}$ u, $\frac{5}{4}$ u to generate new fractional measures of length, and so, too, for iterating by $\frac{1}{8}$ u. Equivalent fractional measures emerge as students come to recognize common locations on a traveled path with different measures. For example, starting at 0, one iteration of $\frac{1}{2}$ u is the same distance from 0 as is two iterations of $\frac{1}{4}$ u or four iterations of $\frac{1}{8}$ u. What is the same is the distance from the origin; what is different is the measure of this distance. This sense of equivalence relies on conceptualizing the same magnitude of length as having varying measures, depending on the unit of measure, a conception developed and revisited in the second and third levels of this learning trajectory.

As in the two previous levels, quantitative comparisons include additive and multiplicative relations. Addition and subtraction are extended to fractional quantities, as in $\frac{4}{8}$ u $+ \frac{9}{8}$ u $= \frac{13}{8}$ u and $\frac{13}{8}$ u $- \frac{9}{8}$ u $= \frac{4}{8}$ u. These sums and differences are tangible as fused lengths and as differences between lengths.

Because multiple measures of the same distance are possible, the earlier noticings of equivalence are repurposed to construct sums with identical units. For example, $\frac{3}{2}$ u + $\frac{3}{8}$ u implies joining or gluing lengths, with distinct units of measure labeled, much as we might report a measure of 4 feet 3 inches. But to report a single measure, we need to employ identical units, so leveraging equivalence, $\frac{4}{8}$ u is the same distance from the origin (0) as is $\frac{1}{2}$ u; $\frac{3}{2}$ u can be re-expressed as three iterations of $\frac{4}{8}$ u, or $\frac{12}{8}$ u; and the expression of joining lengths (or continuing to travel) is now $\frac{12}{8}$ u + $\frac{3}{8}$ u = $\frac{15}{8}$ u, an expression in identical units of $\frac{1}{8}$ u.

Multiplicative relations now include fractional quantities involving splits of 2 and their composition. Students investigate multiplicative relations such as 1 unit is 2 times as long as $\frac{1}{2}$ unit, justifying this conclusion because two copies of $\frac{1}{2}$ u construct 1 u. In addition, some students at this level will understand the reciprocal relation as well: $\frac{1}{2}$ u is $\frac{1}{2}$ times as long as 1 u, so a 2-split of 1 u creates $\frac{1}{2}$ u (Thompson et al., 2014; note that conceiving of this reciprocal relation entails coordinating a common unit length with thinking simultaneously about measuring unit in sub-unit, and sub-unit in unit). Similarly, $\frac{1}{2}$ u is 2 times as long as $\frac{1}{4}$ u, so 2 copies of $\frac{1}{4}$ u construct a length with measure $\frac{2}{4}$ u, which is equivalent to $\frac{1}{2}$ u. And $\frac{1}{4}$ u is $\frac{1}{2}$ times as long as $\frac{1}{2}$ u, so a 2-split of $\frac{1}{2}$ u constructs a length with measure $\frac{1}{4}$ u. However, other multiplicative relations, such as that $\frac{1}{2}$ u is $\frac{1}{3}$ times as long as $\frac{3}{2}$ u, usually elude children who are thinking at this level because the 3-splitting structure has not yet been explored or elaborated. (Students might have seen a colored portion of figure called one-third, but this image does not imply that they grasp the multiplicative relations involved.)

As students become more familiar with symbolizing scales and other representations used in measuring tools like rulers and in marking lengths such as line segments, they also come to understand that the origin of measure, the zero point, is arbitrary. Any number will do as a starting point for measuring, as Figure 3.2 illustrates. For instance, on a ruler, a length of 2 inches can be measured either as the distance from 0 in to 2 in or as the distance from 8 in to 10 in or, for that matter, from $8\frac{1}{2}$ in to $10\frac{1}{2}$ in.

Table 3.4 lists indicators of the development of a theory of measure that encompasses 2-split units, their symbolization on a scale of measure, and reasoning about the zero point.

3-Splitting and Symbolizing 3-Split Units

As students consolidate their understanding of 2-split units and their compositions, they often think about the possibility of a 3-split of a unit length. Three-splitting is a bit more difficult than 2-splitting to construct manually with a paper-strip length, but students again establish the relation between iteration and this partition of the unit: $\frac{1}{3}$ u is the only partition of the unit for which 3 iterations generates the unit length. Students iterate this part-unit to establish new measures, such as $\frac{7}{3}$ u, and composing 3-splits reveals a new structure to the number of congruent parts of a unit produced (3, 9, 27, 81, …). Composing

TABLE 3.4 Indicators of the Conceptual Underpinnings of 2-Split Measure

Description	Examples
Account for change of origin when measurement does not start at zero (whole numbers).	"I can start to measure from the 3 inches on the ruler and take off 3 inches from the result." *Note:* Starting at 2 or 3 is generally more difficult than starting at 1.
Anticipate structure of repeated 2-splits of a unit (2, 4, 8, 16, ... parts).	"If I fold the unit to make $\frac{1}{2}$ of the unit, and then fold it in half again, and then again—3 times—the unit will be split into 8 (equal) parts." (Then folds unit to show that this is true.) "One-half of one-half of one-half makes 8 parts, and another one-half makes 16 parts."
Symbolize multiplicative comparisons involving splits of 2 with words and arithmetic operations.	"$\frac{3}{2}$ u is 3 times as long as $\frac{1}{2}$ u." (Checks by iterating $\frac{1}{2}$ u 3 times.) "$\frac{1}{2}$ u is 2 times as long as $\frac{1}{4}$ u." (Checks by iterating $\frac{1}{4}$ u 2 times.) "1 u is 4 times as long as $\frac{1}{4}$ u." Symbolizes "4 of $\frac{1}{4}$ u is 1 u" as $4 \times \frac{1}{4}u = 1$ u.
Coordinate whole and part units to measure.	"It was 15 inches or $1\frac{1}{4}$ feet." *Note:* These are 2-level units, or units-of-units. "It's 6 quarters long, or we can say $1\frac{1}{2}$ or $1\frac{2}{4}$ units long."
Symbolize relation between origin and partitioned units on scale.	"You don't write in the middle (of the unit). Put it at the end of the part of the unit, so you can see how far you have traveled."
2-split to construct part-unit (2-split part). Iterate part-unit to measure a length. *Note:* Established with $\frac{1}{2}$ u, then with $\frac{1}{4}$ u followed by $\frac{1}{8}$ u.	Student iterates $\frac{1}{2}$ u, aligning endpoint of preceding iteration with beginning point of succeeding iteration. Coordinates iteration with forward number word sequence (one half, two halves, three halves). Student recognizes properties of 2-split part-unit, including equipartition and number of part-unit iterations of equipartition needed to generate the unit length (e.g., 2 iterations of $\frac{1}{2}$ u generate 1 u, 4 iterations of $\frac{1}{4}$ u generate 1 u.)

splits of 2 and 3 extends split-unit measures to new part-units such as $\frac{1}{6}$ u and $\frac{1}{12}$ u. The two new guiding ideas that emerge as a result of these new splitting structures are (1) composing both 2- and 3-splits, and combinations deriving from these compositions, and (2) measurement scale. These new conceptions enable additive and multiplicative comparisons, so that many of the challenges of the previous level are revisited and reconsidered in light of these extensions.

A benchmark lesson (Length unit 8) supports these developments by posing the challenge of re-engineering a standard ruler, but with a scaled version that begins with an unlabeled 24-inch length called a BigFoot. Working with a longer length forestalls merely copying the divisions of unit from a standard ruler and requires students to generate similar subdivisions of unit. Over several days, students generate the unit partitions of a standard ruler by exploring 2-splits, 3-splits, and compositions of 2- and 3-splits to create the divisions of length that are characteristic of a standard ruler. Children discover, for example, that splitting a unit into thirds and then splitting the thirds into fourths creates twelfths. They are often surprised to learn that splitting first into fourths and then into thirds generates the same result. These compositions of splits are again related to scalar multiplication. Briefly, students construct and interpret expressions such as $\frac{1}{3} \times \frac{1}{4}$ BF $= \frac{1}{12}$ BF as reflecting a 3-split of a part-unit length, $\frac{1}{4}$ BF, that results in $\frac{1}{12}$ BF or, perhaps more in line with actual activity, the expression $\frac{1}{3} \times \frac{1}{2} \times \frac{1}{2} \times 1$ BF as corresponding to composing two 2-splits of 1 BF, followed by a 3-split of $\frac{1}{4}$ BF. These ideas about fractional quantities and scalar multiplication may seem challenging for third- and fourth-grade children, but they are given tangibility by students' extensive familiarity at this point with partitioning, accumulating, and traveling. Each "large inch" is one-twelfth of a BigFoot and is further subdivided into halves, fourths, eighths, and sixteenths. Students are challenged to mark these 2-split subdivisions on their BF to mimic standard rulers, and this helps them understand how the designers of rulers rely on hash marks of different vertical lengths to indicate subdivisions of one-half, one-fourth, and one-eighth inch.

Instruction continues to focus on multiplicative and additive comparisons but now includes part-units beyond the variants of 2-splits. For example, $\frac{2}{3}$ u $+ \frac{1}{4}$ u can be re-expressed in common measure as $\frac{8}{12}$ u $+ \frac{3}{12}$ u. Similarly, multiplicative comparisons are further developed, as in $\frac{4}{3}$ Fulk is 2 times as long as $\frac{2}{3}$ Fulk or $\frac{1}{3}$ u is 2 times as long as $\frac{1}{6}$ u, which implies that $\frac{1}{6}$ u is $\frac{1}{2}$ times as long as $\frac{1}{3}$ u.

Table 3.5 summarizes indicators of the conceptual underpinnings of 3-split unit measures and of compositions of 2- and 3-split unit measures.

Generalizing Relationships Among Units and Measures

At the most sophisticated benchmark level in our trajectory, students consolidate and generalize relations that they previously understood only for familiar units and splits of units. For example, given $\frac{1}{n}$ unit, they anticipate that n copies will generate 1 unit. Using knowledge of relations among units, students predict the effects of changes in the unit on measure or scale. For example, if x-units are $\frac{1}{2}$ times as long as y-units, then a measure of 14 y-units will have a measure of 28 x-units. Students can derive relations among units if they are given an expression of the same attribute in different scales of measure.

TABLE 3.5 Indicators of the Conceptual Underpinnings of 3-Split Unit Measures

Description	Examples
Symbolize multiplicative comparisons involving splits of 3 and 2 with words and arithmetic operations.	"1 Goade is 3 times as long as $\frac{1}{3}$ Goade." "$\frac{1}{2}$ inch is 2 times as long as $\frac{3}{12}$ inch." "$\frac{4}{3}$ Fulk is 2 times as long as $\frac{2}{3}$ Fulk." "$\frac{2}{3}$ Fulk is $\frac{1}{2}$ times as long as $\frac{4}{3}$ Fulk."
Account for change of origin when measurement does not start at zero (fractions and whole numbers).	"If I start at 3 and go to $7\frac{1}{4}$, the measure is $4\frac{1}{4}$." "If I travel from $2\frac{1}{2}$ cm to $8\frac{3}{4}$ cm, it's $6\frac{1}{4}$ cm."
Interpret markings on a standard foot ruler in terms of splitting of foot, inch.	Identify inches as $\frac{1}{12}$ ft. Understand how different vertical lengths on a ruler convey 2-splits of inch, such as $\frac{1}{2}$ in, $\frac{1}{4}$ in, $\frac{1}{8}$ in.
Compose splits of 2 and 3 to generate $\frac{1}{6}$ u or $\frac{1}{12}$ u.	"$\frac{1}{2}$ of $\frac{1}{3}$ ft is $\frac{1}{6}$ ft." "$\frac{1}{2}$ of $\frac{1}{2}$ of $\frac{1}{3}$ ft is $\frac{1}{12}$ ft."
Anticipate outcomes of repetitions of 3-split (3, 9, 27, …). Symbolize/write (with words and fraction) relation between origin and 3-split units on scale.	"If you make thirds and then split them again into thirds, you get 9 parts."
Generate a 3-split of a unit, label it as $\frac{1}{3}$ u. Use part-unit ($\frac{1}{3}$ u) iteration to measure a length.	"I split it into 3 parts that are exactly the same." Drags finger, saying, "From here (0) to here (first crease of the split unit) is $\frac{1}{3}$ Goade. And from 0 to here (second crease) is $\frac{2}{3}$ Goade." Iterates to measure a length, $\frac{1}{3}$ u, $\frac{2}{3}$ u, $\frac{3}{3}$ u, $\frac{4}{3}$ u (one third, two thirds, three thirds, four thirds).

For example, "If the measure of the height of the plant is about 10 cm, or about 4 in, then an inch is about $2\frac{1}{2}$ cm." Students flexibly employ ideas about length measurement to invent units as needed or to decide upon reasonable surrogates, such as using time as a stand-in for distance, given knowledge of a particular rate. Students develop more general understandings of reciprocal multiplicative relations between measures of the magnitudes of lengths (cf. Thompson, Carlson, Byerley, & Hatfield, 2014). If length m is $\frac{a}{b}$ times length n, then length n is $\frac{b}{a}$ times as long as length m. If A is $\frac{3}{4}$ B, then $B = \frac{4}{3}$ A. These expressions mean that a length with measure $\frac{1}{3}$ A has the same magnitude as a length with measure $\frac{1}{4}$ B, as illustrated in Figure 3.3.

The conceptual benchmarks that underlie these performances are further described in Table 3.6.

In sum, the levels of benchmark understanding that we have described for length measure span from pre-metric foundations, such as directly

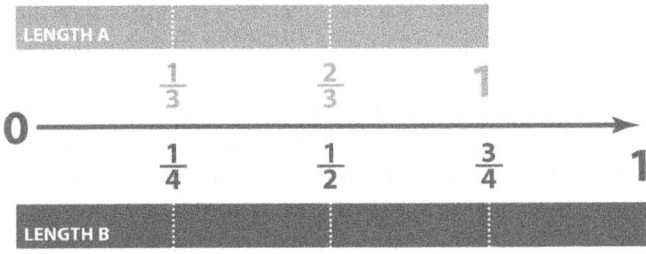

FIGURE 3.3 Reciprocal relations between measured magnitudes of length

comparing objects to decide which is longer and symbolizing the resulting relations, to generalizing and symbolizing relations among differing units and measures. One of the goals of this analysis has been to expose the conceptual complexity and mathematical depth in a domain that is too frequently regarded merely as a set of procedural rules for generating an answer. Clarifying the conceptual underpinnings of length measure shows how measurement can provide a context for motivating and making tangible sense of the broader system of arithmetic operations and rational number (as described in Chapter 7).

TABLE 3.6 Indicators of the Conceptual Underpinnings of Generalizing Relationships Among Units and Measures

Description	Examples
Invent and justify a surrogate measure of length (the unit of measure is not a length).	"We can use the time it takes as a measure of distance, because we can assume a constant rate." "This bushiness index (extent of branching) tells me how much the elodea plant (its total length) grew in the water."
Anticipate reciprocal relation involving multiplicative comparison of the magnitudes of two lengths.	"$\frac{1}{6}$ u is 2 times as long as $\frac{1}{12}$ u. So, $\frac{1}{12}$ u must be $\frac{1}{2}$ times as long as $\frac{1}{6}$ u." "Length A is $\frac{3}{4} \times$ Length B. So, Length B is $\frac{4}{3} \times$ Length A. OK, so if Length B is 20 cm, then length A is 15 cm, and Length B is $\frac{4}{3}$ A because $\frac{1}{3}$ of A is 5 cm, and 4 copies of 5 cm is 20 cm."
Derive relations among units, given expression of the same attribute in different scales of measure.	"If the measure of the height of the plant is about 10 cm or about 4 in, then an inch is about $2\frac{1}{2}$ cm."
Use relations among units to quantify results of changes in unit.	"The measure of the height of the plant is 14 cm. If a cm is ten times as long as a mm, then the measure is 140 mm." "If I change the unit so that it is half as long as the original unit, the measure doubles."
For an unfamiliar split-unit, anticipate that n iterations of $\frac{1}{n}$ unit generate a unit length.	"This length is $\frac{1}{5}$ Goade, so it takes 5 of them to make 1 Goade unit." (Student can literally iterate to check but should be able to predict number of iterations.)

In Chapter 8 we describe selected examples of results of research on students' learning of concepts implicated in length measure. The examples are drawn from the diverse sources of evidence of student learning described in Chapter 2, including task-based clinical interviews, summative assessment items, and formative assessments conducted within participating classrooms. Their purpose is to lend readers a sense of how children, working individually, reasoned with these ideas in a setting apart from the immediate context of instruction. A second source of evidence is constituted by teacher records of student talk, work products, and activity in the immediate context of instruction. These records provide a snapshot of conceptual development across grades and classrooms during the course of a school year.

References

Abelson, H., & diSessa, A. (1981). *Turtle geometry.* Cambridge: MIT Press.

Boulton-Lewis, G. M., Wilss, L. A., & Mutch, S. L. (1996). An analysis of young children's strategies and use of devices for length measurement. *The Journal of Mathematical Behavior, 15*(3), 329–347.

Carey, S. (1985). *Conceptual change in childhood.* Cambridge: The MIT Press.

Carpenter, T. P., & Lewis, R. (1976). The development of the concept of a standard unit of measure in young children. *Journal for Research in Mathematics Education, 7*(1), 53–58.

Clements, D. H. (1999). Teaching length measurement: Research challenges. *School Science and Mathematics, 99*(1), 5–11.

Clements, D. H., & Battista, M. T. (2001). Length, perimeter, area, and volume. In L. S. Grinstein & S. L. Lipsey (Eds.), *Encyclopedia of mathematics education* (pp. 665–668). New York, NY: Routledge Falmer.

Clements, D. H., & Sarama, J. (2004). Learning trajectories in mathematics education. *Mathematical Thinking and Learning, 6*, 81–89.

Clements, D. H., & Sarama, J. (2009). *Learning and teaching early math: The learning trajectories approach.* New York, NY: Routledge.

Clements, D. H., & Stephan, M. (2004). Measurement in pre-k to grade 2 mathematics. In D. H. Clements & J. Sarama (Eds.), *Engaging young children in mathematics: Standards for early childhood mathematics education* (pp. 299–320). Mahwah, NJ: Lawrence Erlbaum Associates.

Confrey, J., Maloney, A., Nguyen, K., Mojica, G., & Myers, M. (2009). Equipartitioning/splitting as a foundation of rational number reasoning using learning trajectories. *Proceedings of the 33rd Conference of the International Group for the Psychology of Mathematics Education, Volume 2* (pp. 345–352). Thessaloniki, Greece.

Confrey, J., Maloney, A. P., Nguyen, K. H., & Rupp, A. A. (2014). Equipartitioning, a foundation for rational number reasoning: Elucidation of a learning trajectory. In A. P. Maloney, J. Confrey, & K. H. Nguyen (Eds.), *Learning over time. Learning trajectories in mathematics education* (pp. 61–96). Charlotte, NC: Information Age Publishing, Inc.

Confrey, J., & Smith, E. (1995). Splitting, covariation, and their role in the development of exponential functions. *Journal for Research in Mathematics Education, 26*(1), 66–86.

Cortina, J. L., Višňovská, J., & Zúñiga, C. (2014). Unit fractions in the context of proportionality: Supporting students' reasoning about the inverse order relationship. *Mathematics Education Research Journal, 26*(1), 79–99.

diSessa, A., & Sherin, B. (1998). What changes in conceptual change? *International Journal of Science Education, 20*(10), 1155–1191.

Ford, M. J. (2006). The game, the pieces, and the players: Generative resources from two instructional portrayals of experimentation. *The Journal of the Learning Sciences, 14*(4), 449–487.

Hiebert, J. (1981). Cognitive development and learning linear measurement. *Journal for Research in Mathematics Education, 12*, 197–211.

Hiebert, J. (1984). Children's mathematics learning: The struggle to link form and understanding. *The Elementary School Journal, 84*(5), 497–513.

Lehrer, R. (2003). Developing understanding of measurement. In J. Kilpatrick, W. G. Martin, & D. E. Schifter (Eds.), *A research companion to principles and standards for school mathematics* (pp. 179–192). Reston, VA: National Council of Teachers of Mathematics.

Lehrer, R., Jacobson, C., Kemeny, V., & Strom, D. (1999). Building on children's intuitions to develop mathematical understanding of space. In E. Fennema & T. A. Romberg (Eds.), *Mathematics classrooms that promote understanding* (pp. 63–87). Mahwah, NJ: Lawrence Erlbaum Associates.

Lehrer, R., Jacobson, C., Thoyre, G., Kemeny, V., Strom, D., Horvath, J., Gance, S., & Koehler, M. (1998). Developing understanding of geometry and space in the primary grades. In R. Lehrer & D. Chazan (Eds.), *Designing learning environments for developing understanding of geometry and space* (pp. 169–200). Mahwah, NJ: Lawrence Erlbaum Associates.

Lehrer, R., Jaslow, L., & Curtis, C. L. (2003). Developing an understanding of measurement in the elementary grades. In D. H. Clements & G. Bright (Eds.), *Learning and teaching measurement* (pp. 100–121). Reston, VA: National Council of Teachers of Mathematics.

Lehrer, R., Jenkins, M., & Osana, H. (1998). Longitudinal study of children's reasoning about space and geometry. In R. Lehrer & D. Chazan (Eds.), *Designing learning environments for developing understanding of geometry and space* (pp. 137–167). Mahwah, NJ: Lawrence Erlbaum Associates.

Lehrer, R., & Pfaff, E. (2011). Designing a learning ecology to support the development of rational number: Blending motion and unit partitioning of length measures. In D. Y. Dai (Ed.), *Design research on learning and thinking in educational settings* (pp. 131–160). New York: Routledge.

Lehrer, R., & Schauble, L. (2015). Learning progressions: The whole world is NOT a stage. *Science Education, 99*(3), 432–437.

McClain, K., Cobb, P., Gravemeijer, K., & Estes, B. (1999). Developing mathematical reasoning within the context of measurement. In L. V. Stiff & F. R. Curci (Eds.), *Developing mathematical reasoning in grades K-12 (1999 Yearbook of the National Council of Teachers of Mathematics)* (pp. 93–106). Reston, VA: National Council of Teachers of Mathematics.

Piaget, J., Inhelder, B., & Szeminska, A. (1960). *The child's conception of geometry.* New York, NY: Basic Books.

Rouse, J. (2015). *Articulating the world.* Chicago: University of Chicago Press.

Simon, M. A. (1995). Reconstructing mathematics pedagogy from a constructivist perspective. *Journal for Research in Mathematics Education, 26*, 114–145.

Simon, M. A., & Tzur, R. (2004). Explicating the role of mathematical tasks in conceptual learning: An elaboration of the hypothetical learning trajectory. *Mathematical Thinking and Learning, 6*(2), 91–104.

Sophian, C. (2007). *The origins of mathematical knowledge in childhood.* Mahwah, NJ: Lawrence Erlbaum Associates.

Szilagyi, J., Clements, D. H., & Sarama, J. (2013). Young children's understanding of length measurement: Evaluating a learning trajectory. *ZDM—The International Journal on Mathematics Education, 44,* 581–620.

Thompson, P. W., Carlson, M. P., Byerley, C., & Hatfield, N. (2014). Schemes for thinking with magnitudes: A hypothesis about foundational reasoning abilities in algebra. In K. C. Moore, L. P. Steffe, & L. L. Hatfield (Eds.), *Epistemic algebra students: Emerging models of students' algebraic knowing.* WISDOMe Monographs (Vol. 4, pp. 1–24). Laramie, WY: University of Wyoming.

Van Fraassen, B. C. (2008). *Scientific representation: Paradoxes of perspective.* New York, NY: Oxford University Press.

4

CREATING NEW QUANTITIES IN THE DYNAMIC GENERATION OF AREA

In this chapter we describe continuity and changes in students' conceptions of area and its measure. The description is again organized as a progression of networks of ideas and relations that emerge as students participate in instruction designed to provide opportunities for investigating this realm of measure.

Two Coordinated Perspectives on Measure

Instruction supports learning by helping students conceive of area measure from a dual perspective. The first perspective arises as students construct and reflect on properties of units that can be iterated to compare magnitudes of areas in ways that are reliable and easily adopted by others. These ideas are conceptually parallel to those that underlie the iterative unit measure of length. However, in the context of area measure, iteration is now adapted to the challenges posed by an additional dimension of measured space. For example, students often find it challenging to structure an area into units by coordinating corresponding units of length to dissect a shape into units that can be counted to generate a quantity of area (Battista, 2004; Battista, Clements, Arnoff, Battista, & Van Auken Borrow, 1998; Outhred & Mitchelmore, 2000). Moreover, they often place unconventional constraints on units of area measure, choosing only those that resemble the figure of the unknown area or those that can be contained within its boundaries (Lehrer, Jenkins, & Osana, 1998).

The second perspective stems from visualizing area and unit dynamically, as a product of the movement of one length through another at different angles (Kobiela & Lehrer, 2019; Lehrer, 2003). This dynamic perspective grows from thinking about length as movement. Now, however, movement of one length through another generates two-dimensional space. As we soon describe more

DOI: 10.4324/9781003287476-4

completely, the first perspective, constructing unit iteration, trends toward structuring areas, especially of simple convex figures, as arrays of composite units (e.g., rows and/or columns of units in rectangular areas), primarily so that they can be more efficiently counted (Cullen et al., 2018). In contrast, generating area (and unit) dynamically links area more explicitly to multiplication, as the product of lengths (Thompson, 2000, 2011). This second perspective trends toward thinking of area as a rate of generation of area as one measured length is moved through another measured length. For instance, moving 3 inches through 6 inches generates 3 square units of area per inch moved through the 6-inch length (Brady & Lehrer, 2021). Integrating the iterative and dynamic perspectives on area positions children to construct and justify formulas for familiar polygons, such as rectangles, squares, parallelograms, and regular hexagons, with extension to circles (Lehrer & Slovin, 2014). The interplay of both perspectives is articulated as a model trajectory initiated in everyday conceptions and experiences of covering.

Integrating the Two Perspectives on Measure

Direct Comparison of Magnitudes

Area is interpreted intuitively by young students as covering or filling a portion of a planar region (Lehrer, Jenkins, & Osana, 1998). Magnitude of area refers to the amount or extent of cover. Area as space covered within a region of the plane is consistent with a metaphoric extension of experience with containers, in which boundaries of the container determine spatial relations such as inside and outside (Lakoff & Núñez, 2000). Wall coverings, such as tapestries, and other artifacts from the designed world, such as painted walls and tiled floors, exemplify enclosed space. If one planar region encloses the other or if the regions are directly congruent (e.g., a motion can bring a figure denoting a bounded region of a plane directly onto the second figure), children will typically feel confident in judging the regions, including 2-D figures, as covering less space, more space, or the same amount of space (Lehrer, Jacobson, et al., 1998). Symbolic expression of relations among magnitudes of space covered can be initiated with these first understandings, as in Area (A) < Area (B), Area (A) = Area (B), or Area (A) > Area (B). This constitutes the first, entrée level of the trajectory.

Comparing Magnitudes of Area Indirectly Through Dissection and Unit Dissection

Some approaches to area in the elementary grades advise the use of units to cover and count figures to compare magnitudes of area. Unfortunately, this approach typically does not lead to insights about how these units are constructed or help students grasp their properties (Smith, Males, & Gonulates, 2016; Zacharos, 2006). To bring these understandings to light, teachers challenge students to dissect different-looking rectangles to compare the magnitudes of their areas (which covers the most space?) and, in the process, to adopt a common dissected part as a unit of measure (see Area unit 1).

For example, second-grade children decided which of three paper rectangles covered the most space by folding and rearranging partitions (dissecting the area) to establish relations among the magnitudes of their areas (Lehrer, Jacobson et al., 1998; Strom, Kemeny, Lehrer, & Forman, 2001). The rectangles were perceptually distinctive, with dimensions of 1 u × 12 u, 2 u × 6 u, and 4 u × 3 u, so most children expressed the belief that the rectangles covered different amounts of space, especially the "long, skinny" rectangle (that is, the 1 × 12). Students could not bring the rectangles into direct congruence with isometries (e.g., through reflection or rotation), and they were not provided rulers to mark units of length or tangible units to cover and count the space contained by each rectangle. Consequently, dissecting (by paper folding) and rearranging parts of rectangles emerged as a practical avenue for comparing areas. Figure 4.1 depicts a dissection, proposed by a student, that established

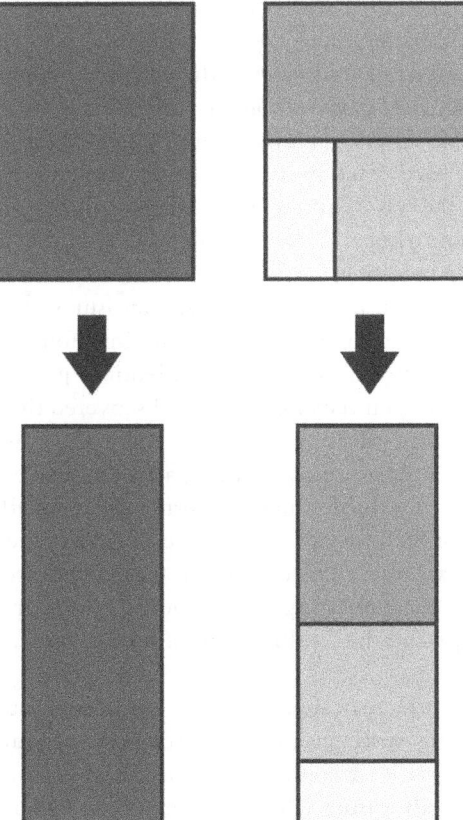

FIGURE 4.1   Establishing congruence of spaces covered by dissection

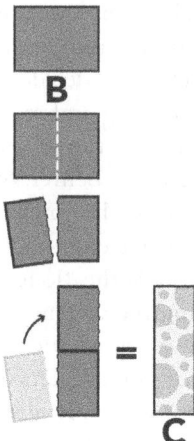

FIGURE 4.2 Congruence of areas of different rectangles by identical-part dissection

equivalence in amount of space covered between 4 u × 3 u and 6 u × 2 u rectangles (recall that dimensions were not labeled).

As students generated and shared dissections, many struck upon the notion of using identical parts to compare magnitudes of area, as illustrated in Figure 4.2. In this instance, a few children dissected the 3 u × 4 u rectangle into two identical parts, each measuring 3 u × 2 u, and then rearranged the parts to construct the 6 u × 2 u rectangle. Identical parts provided a pathway for thinking of comparing areas by referring to a common unit.

Students invented multiple variations of identical-part dissections and came to recognize those identical parts as enabling measures of area. For example, several students split the 1 u × 12 u rectangle into four identical 1 u × 3 u parts and subsequently rearranged these identical parts to constitute the 3 u × 4 u and the 2 u × 6 u rectangles. They discovered that each of the three rectangles had the same measure of four of this 1 u × 3 u rectangular unit. Other students invented a square unit by partitioning the 3 u × 4 u rectangle first horizontally by thirds and then vertically by fourths, constructing a unit square, as illustrated in Figure 4.3. They found that each rectangle had a measure of 12 square units. The upshot was that students could compare areas of the three rectangles by referring to unit measures, which varied with choice of dissection strategy and the resulting unit of measure (e.g., 4 of 1 u × 3 u and 12 of 1 u × 1 u).

Children went on to rearrange 12 of the unit squares to generate figures that were perceptually distinct but all had the same measure and, hence, the same magnitude of space covered. Conserving area as a quantity is one of the core concepts of area measure (Piaget, Inhelder, & Szeminska, 1960; Smith et al., 2016), and student engagement with partitioning and rearrangements of partitioned space afforded practical experience of this conception.

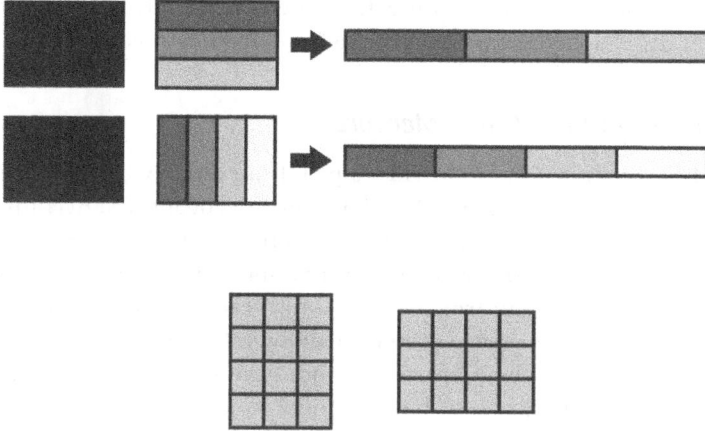

FIGURE 4.3 A square unit dissection produced by a sequence of horizontal and vertical partitions of the 3 u × 4 u rectangle

Dissection embarked students on the construction of unit, an achievement that permitted indirect comparison of magnitudes of area. Since students constructed units of measure to solve a practical problem of comparison, the role of unit in establishing indirect measure emerged during the course of the activity, rather than being prescribed *a priori*. In short, students had the opportunity to experience the problem of measure for which unit is a solution. Moreover, students' invented units included rectangles of different dimensions as well as a square, an outcome that avoided rote emphasis on only square units. These roles for dissection are depicted by the performance indicators listed in Table 4.1, which is excerpted from the complete area construct. Dissection based on identical

TABLE 4.1 Indicators of Conceptual Underpinnings of Unit, Level 2 of Area Trajectory

Description	Examples
Dissect a shape into identical, congruent parts. Use a part as a unit of measure.	Comparing a 1 u × 12 u rectangle to a 4 u × 3 u rectangle: "I folded the paper to make 4 of these (gestures to 1 u × 3 u partition) and then I used all 4 of them to make this one (the 1 u × 12 u). So, they cover the same amount of space."
Dissect (partition) a shape into parts and rearrange the parts to compare areas. (The parts are not congruent.) *Note: This implies that child understands that area is unaltered by partitioning a shape and rearranging its parts (conservation).*	"I cut this shape into different parts and then rearranged them like this—see, I made this other shape without any leftovers—so I know that the two shapes cover the same amount of space."

partitions is often preceded by dissection based on non-identical partitions, so that unit dissections are capstones of this level of the area construct.

Properties of Units of Area Measure

As we have seen, dissection provides a path from directly comparing magnitudes of area to comparing areas based on measured quantities. Accordingly, in the next, third level of the progression, properties of units of area measure are more thoroughly interrogated to establish reliable measure. As in length measure, we aim to arrange instruction so that the need for shared understandings of unit emerges in the course of student activity. Common properties of unit in area measure have counterparts in length measure—for instance, the need for units to tile the space, to fracture units to achieve greater precision in measure, and to anticipate effects of changes in unit on the resulting measure. Other properties arise from the challenges posed by a second dimension of length, including differentiating units of length measure from those of area measure and coordinating intervals of length to dissect figures into identical unit arrays (e.g., Battista, 2004; Outhred & Mitchelmore, 2000). Indicators of these conceptual underpinnings are outlined in Table 4.2 and are arranged so that earlier developing conceptions of unit properties are located at lower portions of the table.

To promote investigation of the properties of unit that expedite reliable measure, we typically invite students to compare the magnitudes of areas of nonpolygonal figures, like handprints or leaves. Measure is a practical avenue for this comparison, particularly when students realize that strategies like direct congruence are not feasible (handprints are difficult to bring into direct correspondence). Students choose among several potential units, typically including dried beans, strands of spaghetti, coins, and inch- and centimeter-squared grid paper. Solutions like those displayed in Figure 4.4 are commonplace and reflect student beliefs that units of area measure should resemble the figure being measured (e.g., follow the contour of the figure) and/or that units must be contained within the bounded space (Lehrer, Jenkins, & Osana, 1998).

When second-grade students compared the measures of the same handprint with units like these, they realized that the resulting measures were not reliable. Students obtained different measures of the area of the same handprint, and individual students noticed that the measure of the same handprint was not consistent from one time to the next (Lehrer, Jacobson et al., 1998; Lehrer, Jaslow & Curtis, 2003). As these inconsistencies in measure were highlighted during class conversations, students began to realize that discrepancies in measure arose because the sizes of the same kinds of beans varied (identical unit property) and because their choice of unit did not tessellate the space, leaving regions of space unmeasured. These problems in reliable

TABLE 4.2 Indicators of Conceptual Underpinning of Properties of Unit

Description	*Examples*
Interpret area measure as ratio of measured area to unit or to composite unit of measure.	"The area is 12 square units, which is 12 times the area of 1 square unit." "The area measure of each column of the 3 u × 6 u rectangle is 3 u², so the area of the rectangle (18 u²) is six times the area of the column."
Dissect shape into units by coordinating intervals of length.	"This rectangle is 3 in × 4 in. So here are the 12 square inches of the area measure."

4 inches

Differentiate units of length (e.g., perimeter) and area measure.	"This line segment does not have area because it represents the distance between these two points. Its thickness does not matter—it's like a sidewalk; it can be wider or narrower but it's the distance between (landmarks) that matters." (Paraphrasing of classroom conversation) "Perimeter is how long, so we use inches, but area is space inside, so we use square inches."
Anticipate inverse relation between area of unit and measure of area.	"The area measure is greater with smaller square units, because you can fit more inside."
Recognize and justify closure of figure as necessary for area measure.	"If it (any figure) has an opening, the area leaks out, so you just have to keep counting and counting, so you will never stop."
Partition unit to tile space. Combine whole-number and fractional units to establish a measure.	"I couldn't use whole units for these parts of the figure, so I broke the square unit into $\frac{1}{2}$ and $\frac{1}{4}$ parts. Then I found the area as $5 + \frac{1}{2} + \frac{1}{2} + \frac{1}{4}$. That's $6\frac{1}{4}$ of these squares to cover it."
Tile an area with identical units and justify. (Or, if different units are used, label them as such.)	"The units should touch so there is no empty space. You can't measure what you don't cover." "The beans did not work to measure the area of my hand. They left spaces. And they weren't all the same size, so I did not get the same number each time I tried to fit them into my handprint."

measure were resolved by employing identical units (e.g., inch-square grid paper), as in Figure 4.4b, but this solution also generated a new challenge of accounting for partial units. The solution shown in Figure 4.4b approximated whole units by matching parts that students judged could be composed to constitute a whole unit (students used matching colors to mark parts that

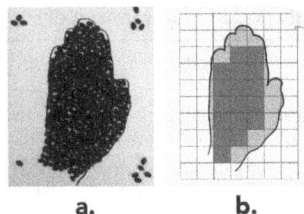

a.   b.

FIGURE 4.4(a) and (b) Two student solutions to measuring the area of a handprint

collectively constituted a whole unit, although the black-and-white figure used here does not showcase this strategy). Other students quantified fractional parts of the square unit, such as $\frac{1}{2}$, $\frac{1}{4}$, and $\frac{1}{3}$, which were then combined to generate an approximate measure (Lehrer, Jacobson et al., 1998). By comparing solutions with square-cm vs. square-in grid paper, students also considered the effects of choice of unit on the measure of a handprint's magnitude (e.g., a measure of $10\frac{1}{2}$ square inches vs. approximately 68 square centimeters). Many of these student experiences of how properties of unit support reliable measure in area have counterparts in their previous experiences with length measure, so these commonalities can be highlighted with appropriate support from teachers to contribute to a growing theory of measure—a set of understandings and performances that constitute anticipations of how measures ought to be constructed and how they ought to behave.

As students grapple with ways of getting a grip on units that are useful for measure, they also revisit their intuitions about enclosure, but now as a formal property of measure. To stimulate reflection about what was formerly understood intuitively, teachers often pose contrasting examples of figures without closure, such as the pair depicted in Figure 4.5.

Students usually argue that the form on the left is "bigger" but quickly come to see that because the figure is not closed, the count of unit is unbounded. For instance, several third-grade students proposed that the figures "leaked" into the surround, so the surround would have to be tessellated with units, but

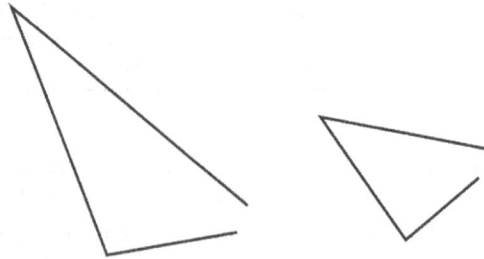

FIGURE 4.5 Contrasting cases of non-closed forms

students were concerned that they could not identify any agreed-upon way of stopping the process. Consequently, both figures have the same unknowable measure. In this way the property of closure emerged as a sensible constraint arising from acts of measure, rather than as an arbitrary property to be memorized.

The process of dissecting figures into units is extended at this succeeding level to include the generation of unit dissection by coordinating intervals of unit length, including composites of units (Cullen & Barrett, 2020; Cullen et al., 2018). These acts of coordination also support differentiation of the units of length used to measure perimeter from the units of area used to quantify area (see Area unit 4). For example, within the context of designing animal enclosures for a local zoo they had recently visited, second-grade students contrasted the areas and perimeters of rectangles and of polygons composed of rectangles (Lehrer, Jacobson et al., 1998). Different enclosure designs, as shown in Figure 4.6, were presented on paper and scaled in inches, with the scaling factor chosen at the discretion of the teacher (e.g., 1 inch = 1 yard, 1 square inch = 1 square yard). The second-grade students did not know a formula for the area of a rectangle but had access to standard rulers. The teacher asked students to show units of area measure, beginning with a 5 in × 8 in rectangle and a 4 in × 10 in rectangle. In subsequent versions of this task, students drew a line just as long as the perimeter of each figure to foster visualization of perimeter as length.

To compare the 5 in × 8 in and 4 in × 10 in enclosures, students most often marked unit lengths on one vertical and one horizontal side of the 5 in × 8 in rectangle (Lehrer, Jacobson et al., 1998; Lehrer, Jaslow, & Curtis, 2003). Then they applied a strategy like the one they had invented to compare the area of rectangles by folding. That is, they drew lines horizontally in parallel and then vertically in parallel to dissect the rectangle into unit squares, an invented strategy that appears to be generally effective at promoting unit dissection (Cullen et al., 2018). To find the area, some students proposed counting 8 rows of 5 square units, whereas others counted 5 columns of 8 square units. Only a few counted all the units one by one. By skip counting the composite unit, the students established that the measure of area was 40 in² and that the rectangle could be structured as composite units, either of 5 square units or of 8 square units, or more simply (and less efficiently) as an array of 40 square units. Students proceeded in a similar fashion with the 4 in × 10 in rectangle to find again that rectangles with different dimensions could have the same

A B C D E F

FIGURE 4.6 Comparing perimeter and area of different compositions of rectangles

area measure (the 4 in × 10 in rectangle also had an area of 40 in^2). Yet students also noted that in spite of their identical area measure, the enclosures did not require the same amount of fencing. In other elaborations of this lesson (see Area unit 4), students used masking tape on the desk or floor to create lengths as long as the perimeter, an action that highlights perimeter as a length. Animating the path of the enclosure as a walk additionally helps students focus on the distinction between perimeter and area. In the second-grade classroom, students eventually proposed that a more efficient strategy to find the area measure of rectangles would be to multiply length and width, and the teacher agreed, with the provision that students needed to be able to show the units of area measure. (This student conception regards product as repeated addition, based on the iteration of units or composite units, a concept like that of measuring a length as a number of copies of a unit length.)

The second-grade students went on to dissect figures like those in items B, C, and E in Figure 4.6 to determine area and perimeter measures. Most dissected the figures into rectangles or squares, determined the area measure of each dissected part, and then summed the measures to determine the measure of the figure. A few students employed a subtractive dissection strategy. For example, Figure B's area measure was found by subtracting the area of an imagined rectangle in the upper right corner (4 in vertically × 3 in horizontally) from an imagined larger rectangle (8 in × 6 in), which was identified by extending the horizontal and vertical sides of the figure. In summary, the procedures of dissection, which were started at the outset of the progression to support initiation into inventing units of area measure, were redeployed to accomplish new insights about relations between intervals of length and structuring of rectangles by unit of area measure or by composites of unit. These procedures further supported the development of strategies for measuring the area of more complex polygons as composites of rectangles and squares.

Structuring the area measure of rectangles via units and composites of units also supports multiplicative comparison (see Area unit 4) between measures of area of a figure and unit or composite unit. For example, the rectangle with side lengths of 5 u and 8 u invites statements like "40 in^2 is 40 times as much space covered as is 1 in^2 (40 iterations of 1 in^2)" or "... is eight times as much space covered as is 5 in^2 (8 iterations of 5 in^2)" or "... is five times as much space covered as is 8 in^2." Writing an expression for each of these relationships, as in 40 in^2 = 5 × 8 in^2, anchors tangible measure of space to multiplicative comparison.

Dynamic Generation of Area and Product

After students have considered area and its measure from the unit-iterative perspective, they are introduced to a complementary perspective, visualizing area dynamically as a continuous movement or sweep of one length

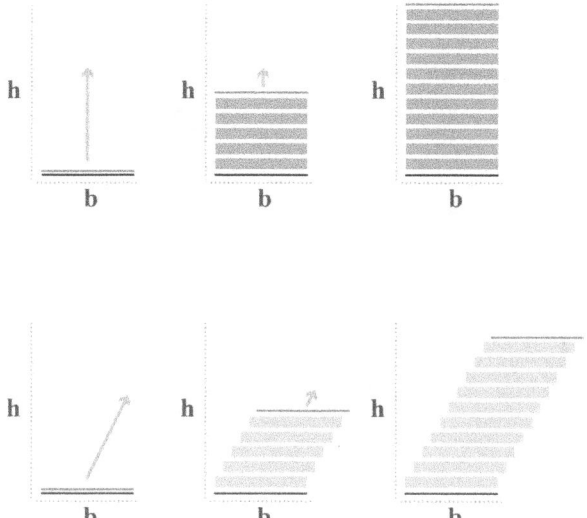

FIGURE 4.7   Area is generated dynamically by movement of one length through another

through another (Lehrer, 2003; Lehrer & Slovin, 2014). When one length is swept perpendicular to another, the result is a rectangular area. When one length intersects another at an angle (other than 0 or 180 degrees), other figures are possible, as illustrated in Figure 4.7 (the white spaces in the figure are intended to help readers envision the movement through units of length, but they should be understood as continuous movement through the plane). Units are generated by the same process—that is, sweeping one unit length through another, as in the generation of a square foot by moving one foot length through another foot length oriented at a right angle to the first.

The rate of generation of area anchors the multiplication of lengths to a continuously emerging product (Brady & Lehrer, 2021). For example, as illustrated in Figure 4.8, moving a 4-in length through 5 in at a right angle allows students to literally see that for every inch swept of the inch length, 4 square inches are produced. Moreover, for every half-inch swept of the 5-in length by the 4-in length, 2 square inches are produced, and for every one-fourth inch swept, 1 square inch is produced. Third- and fourth-grade students can redeploy the 2-splitting structure they explored during the measure of length to help them understand that the quantity of area produced can be arbitrarily small but is nonetheless continuously generated as one length is moved through another. The multiplicative comparison of quantity generated supports this rate-based conception of product in that each succeeding 2-split results in a product either one-half or two times that of its predecessor.

Further elaboration of multiplicative product is promoted by sweeping lengths with different units of measure (Kobiela & Lehrer, 2019; Thompson, 2000).

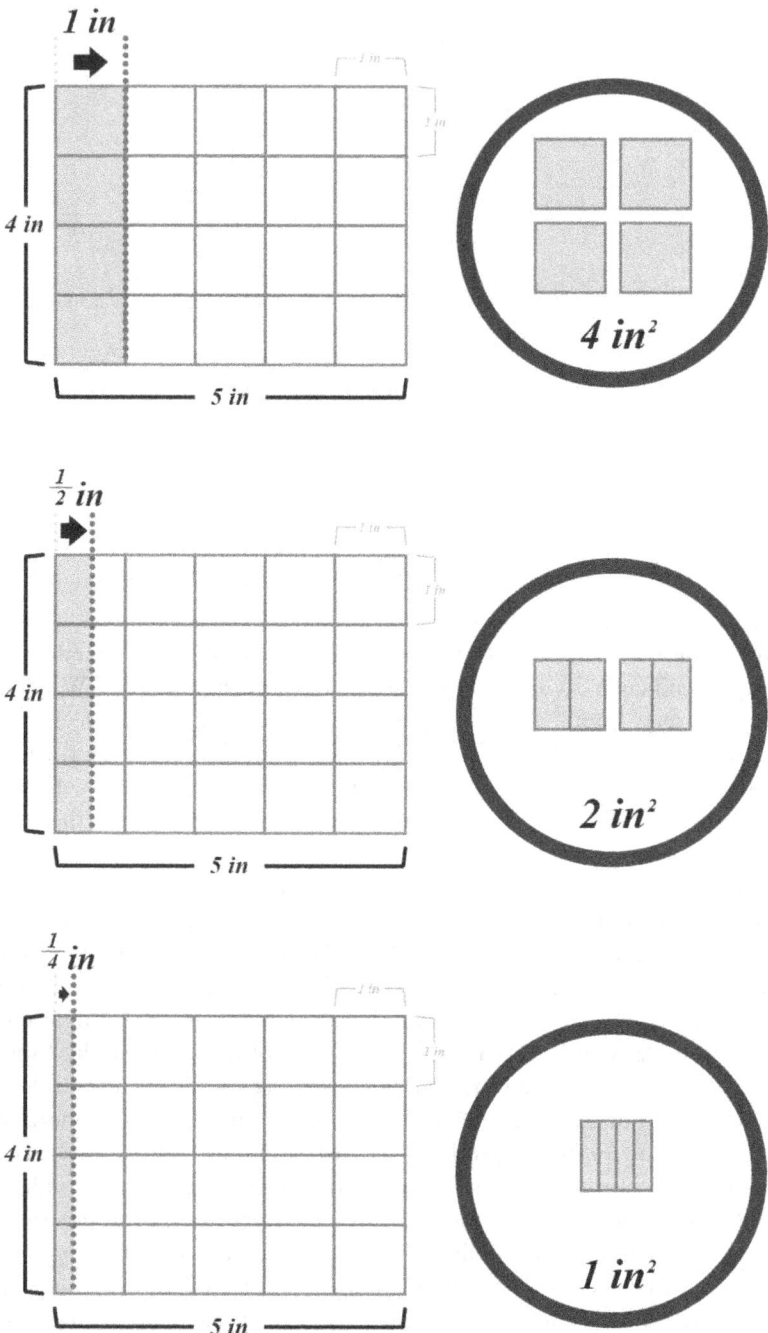

FIGURE 4.8 Sweeping generates measured area per unit length measure

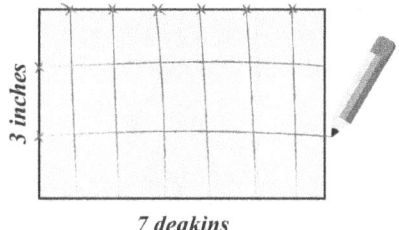

FIGURE 4.9 Sweeping with lengths measured in different units

For example, Figure 4.9 shows that moving a 3-in length horizontally through a 7-deakin length generates 3 in-deakin per deakin unit. The resulting unit dissection is obtained by coordinating the different unit lengths. The magnitude of the area is 21 times the magnitude of an in-deakin, and the perimeter of the figure is 6 in + 14 deakin.

The dynamic generation of area supports geometric images of arithmetic operations and properties of these operations. For example, a 4 in × 5 in rectangle can be generated either by sweeping a 4-in length through a 5-in length or by sweeping a 5-in length through a 4-in length (at a 90° angle in each case). Hence, sweeping establishes that this form of multiplication is commutative, as displayed in Figure 4.10 (Kobiela & Lehrer, 2019).

A further example of geometric images of arithmetic operations is illustrated in Figure 4.11, in which the distributive property of multiplication over addition is modeled as the product obtained by sweeping 4 in through 3 in and then continuing to sweep through another 5 in, expressed symbolically as 4 in × (3 in + 5 in) (Kobiela & Lehrer, 2019).

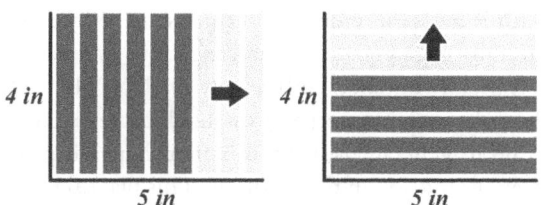

FIGURE 4.10 Commutative property of multiplication

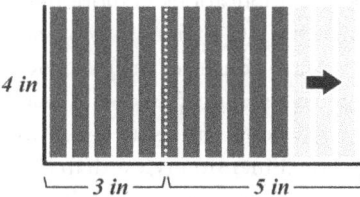

FIGURE 4.11 Modeling the distributive property dynamically

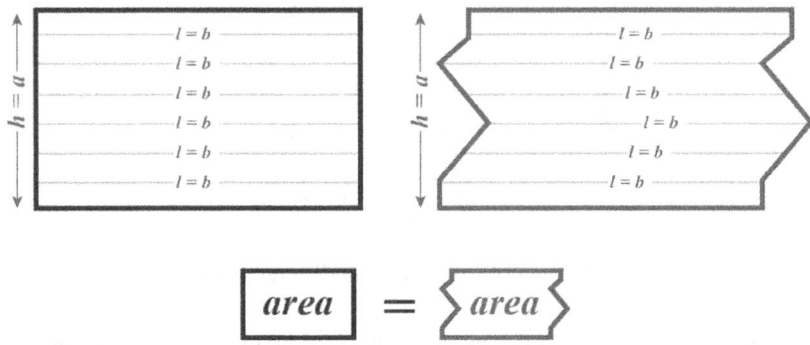

FIGURE 4.12   Cavalieri's principle of equivalent areas

A generalization of sweeping, Cavalieri's principle, states that the areas of figures with identical bases and heights are of equal measure, in spite of differences in their appearance. For instance, in Figure 4.12, the base length is moved directly through the height on the left, but with an oscillating horizontal (but parallel) motion on the right. The areas of the figures are the same, a statement that can be verified by dissection.

Table 4.3 summarizes the network of concepts that we have been describing, again arranged with initial conceptualizations of dynamic area in the lower portion of the table and elaborations and implications of these initial understandings in the upper region of the table. These are concepts and their interrelations that usually emerge as students are inducted into a dynamic perspective on the generation and measure of area.

Classroom tasks and tools to promote the development of a dynamic perspective on area and its measure are described in detail in the project instructional materials (unit 3, Area). Here we outline some of the cornerstone tasks and tools employed to stage productive investigations of area dynamics.

Investigation is usually initiated with readily available physical tools: a squeegee, ceramic tile, rulers, and fingerpaint or shaving cream. Students are challenged to find ways to generate as many different shapes as possible, but doing so through continuous motion in the medium. This playful initiation helps students understand that many different forms are possible, including familiar ones like parallelograms, rectangles, and circles. Proprioceptive and visual feedback helps students understand the implications in motion for some figures, like parallelograms, where it is necessary to maintain a constant angle of the squeegee to the base of the tile in order to construct the parallelogram (the "swept" area).

The length of a squeegee is generally 5 in or 8 in, and students are challenged to construct rectangles (by sweeping the squeegee through the fingerpaint or shaving cream) in which one dimension is less than the length of the squeegee. This task highlights the need to coordinate two dimensions of length. Once a figure has been "swept," students use thin spaghetti to create unit dissections, such as square units, for squares, rectangles, and parallelograms.

TABLE 4.3 Indicators of the Conceptual Underpinnings of a Dynamic Perspective

Description	Examples
Use Cavalieri's principle to justify informally why the areas of figures with identical bases and heights are of equal measure, despite differences in appearance.	"When I swept this length (gestures to 6-in base of figure) and then moved it (gestures parallel to base) back and forth through 5 inches, it has the same area as this one (6 in × 5 in rectangle) because it (gestures to figure produced by back and forth) is just the same space pushed in and out. The push in and out balances.
Extend product conception of area to include finding and comparing areas with dimensions given in unlike units (e.g., length in cm, width in inches).	"If the rectangle is 6 inches long and 4 centimeters wide, the area is 24 inch-centimeters—because I sweep an inch through a centimeter to make an inch-centimeter."
Extend conception and symbolization of dynamic product to include fractional lengths.	"$\frac{1}{2}$ in sweep through $\frac{1}{2}$ in is $\frac{1}{4}$ in^2."
Relate images of sweeping and actions of dissection to symbolic expressions of arithmetic operations and properties.	Sweeping 3 in through 5 in makes the same area as sweeping 5 in through 3 in, so order does not matter: 3 in × 5 in = 5 in × 3 in." (commutative property) "Sweeping 3 in through 5 in and then keep sweeping another 2 inches means 3 in × (5 in + 2 in) = 15 in^2 + 6 in^2." (distributive property of multiplication over addition)
Imagine area as generated by motion of lengths.	Generate areas by sweeping one length through another at a variety of angles (excluding 0° and 180°). (See Figure 4.7 in text.)
Relate motion to product formula for rectangle ($l \times w$ means sweep l through w). Dynamically generate unit of area measure.	"If this rectangle is 6 inches long and I sweep that through the width of 4 inches, then it would be 6 in × 4 in = 24 square units." "I can make a square inch by moving the sweeper 1 in through another inch."

As students move one length through another, as in a 4 in × 8 in rectangle, they are asked to consider how much area is generated per unit of length. Doing so helps students think of composite units, and teachers often raise the conceptual ante by asking students to consider the area generated when a length is swept through a fractional unit of another length, such as by increments of $\frac{1}{2}$ in. Students generate dissections of rectangles with unlike units by, for example, taking the length of the squeegee as one unit and an inch as the other unit, so that different measures of the same rectangle can be compared (e.g., in^2 vs. squeegee-in*). Using unlike units also highlights how a product arises from the motion of one length through another, whatever the respective units of measure (Thompson, 2000). Motions are expressed symbolically, and students use symbolic expressions to generate the corresponding products dynamically (e.g., 5 in × 5 in as 5 in swept through 5 in), so that symbolic expression and motion eventually become exchangeable. Multiplicative comparisons of area measures are revisited. For example, sweeping 5 in through 6 in generates an area with measure 30 in^2, a quantity that is 30 times as much as 1 in^2 and 6 times as much as the composite unit of 5 in^2 that is produced per inch (through the 6-in length). Students also explore Cavalieri's principle informally by moving the same horizontal length through the same vertical length to produce different-looking figures with the same area, as shown previously in Figure 4.12.

After these initial explorations, students employ a digital tool that facilitates the generation and dissection of areas (Brady & Lehrer, 2021). The digital tool, designed by C. Brady, employs a stick-like length that students readily think of as being like a squeegee, and the tool also allows students to specify different units and subdivisions of units. Perhaps most importantly, students can perform both unit dissections and other dissections on swept areas, so that generation and dissection can be coordinated as tools for thought. Figure 4.13 displays a third-grade student's construction of a parallelogram, followed by

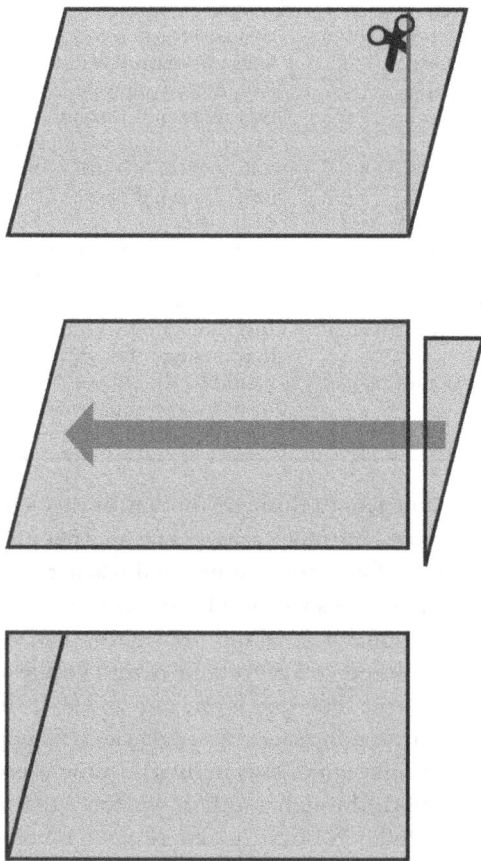

FIGURE 4.13 Swept parallelogram cut and dissected into rectangle

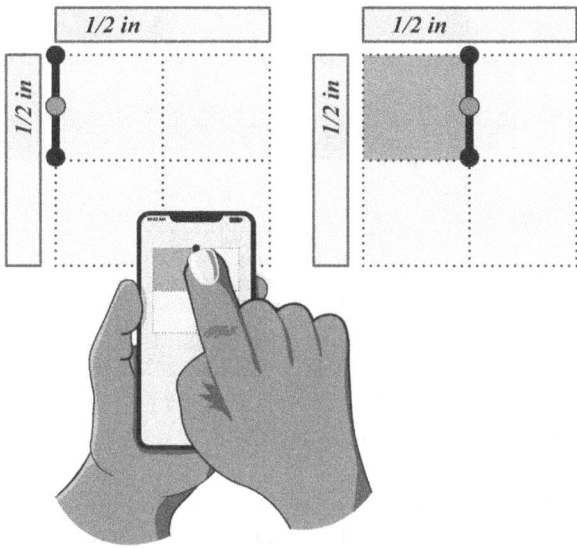

FIGURE 4.14   Sweeping $\frac{1}{2}$ in through $\frac{1}{2}$ in

its dissection into a rectangle to establish the equivalence of area measures of a parallelogram and rectangle with the same base, and where the height (the vertical cut) of the parallelogram is of the same measure as the corresponding side length of the rectangle.

In Figure 4.14, a third-grade student used the subdivision and enlargement tools in the application to investigate $\frac{1}{2}$ u $\times$ $\frac{1}{2}$ u as moving one length through another to generate $\frac{1}{4}$ u^2. Other third-grade students found multiple solutions for generating $\frac{1}{4}$ u^2, such as moving $\frac{1}{4}$ u through 1 u.

Guided Reinvention of Area Measure Formulas

Armed with conceptual tools of dissection and dynamic generation of area by moving lengths, students are now in a position to reinvent formulas for area measure of familiar polygons. The instructional means to do so are outlined in Area unit 5 in the instructional materials. Here dissection and motion are resources for inventing formulas, assuming that students are now familiar with the measure of a rectangle's area as the product of the measures of the lengths of its sides. To construct a formula for the parallelogram, two avenues can be pursued. The first is to use Cavalieri's principle to verify that the areas of a parallelogram and a rectangle of the same height (call one side of the rectangle its height) and same length base must be the same. The second avenue uses dissection, as in Figure 4.13, to establish equivalence between parallelograms and rectangles that share the same base and height. Hence, one can say that the area of a parallelogram is simply the product of its base and height.

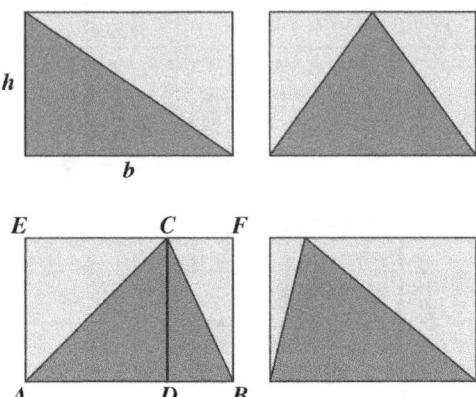

FIGURE 4.15 Justifying area measure of a triangle

Cavalieri's principle and dissection are also resources for generating the area formula of a triangle (Lehrer & Slovin, 2014). For instance, in Figure 4.15, the upper left panel depicts a triangle that clearly occupies half the area of the rectangle, so its area is $\frac{1}{2}$ the product of base and height (any side of the triangle can serve as the base, each with a corresponding height). Then, by Cavalieri's principle, since triangles like the one depicted in the upper right panel share a common base and height with the triangle depicted in the upper left panel, they must all be of the same area (as can be seen by matching congruent widths, which need not be on the same parallel) and hence have the same measure of $\frac{1}{2}(b \times h)$. This will be true even if the height of the triangle is extended beyond the confines of the rectangle.

Another approach relies on dissection to show that every triangle in Figure 4.15 has the same half-area of the corresponding rectangle. For example, in the lower left panel of Figure 4.15, Area $\triangle ABC$ = Area $\triangle ADC$ + Area $\triangle CBD$. The area of $\triangle ADC = \frac{1}{2}$ Area AECD, and the area of $\triangle CDB = \frac{1}{2}$ Area DCFB. Hence, Area $\triangle ABC = \frac{1}{2}$ Area $AEFB = \frac{1}{2}b \times h$. Cases of triangle not enclosed in a rectangle, as displayed in Figure 4.16, are nonetheless amenable to dissection. Referring to this figure and considering the triangle labeled ABC, Area $\triangle ABC$ = Area $\triangle ADC$ − Area $\triangle BCD$. This difference is $\frac{1}{2}$ Area $(AECD) - \frac{1}{2}$ Area $(DCFB)$, which is the same as $\frac{1}{2}$ Area $AEFB$ and, therefore, Area $\triangle ABC = \frac{1}{2}b \times h$.

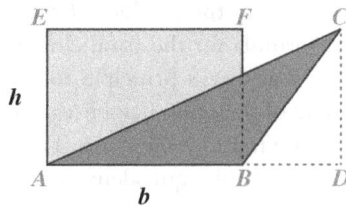

FIGURE 4.16 Visualizing area measure of a triangle not enclosed in a rectangle

Armed with this cluster of formulas for the area measure of a rectangle, parallelogram, and triangle, and apprehending the network of conceptual relations that emerges during the construction of these formulas, students are in a position to generate (with supporting instruction) formulas for the area measure of regular polygons and circles (see unit 5, Area). Performance indicators of the conceptual attainments of this last level of the area progression are illustrated in Table 4.4.

In summary, the growth in conceptions of area and its measure that is outlined in this portion of the measurement progression is initially catalyzed by commonplace intuitions about area as space covered and space contained, along with ideas that space covered does not change when it is partitioned and rearranged. The latter notion is often regarded as an instance of mental conservation that is a barrier to more formal conceptions of area (Piaget & Inhelder, 1969), but simple materials like paper and actions like folding, cutting, and rearranging are

TABLE 4.4 Indicators of Conceptual Underpinnings of Area Formulas

Description	Examples
Characterize circle as the result of a limiting process (infinitely sided n-gon) and generalize area formula for n-gon to generate and justify formula for measure of area of a circle.	"The perimeter of the circle is $\pi \times 2r$. That's true because π is the ratio of the length of its circumference (its perimeter) to the length of its diameter. If you think of a circle as having many, many small sides, its area is $\frac{1}{2} \times perimeter \times h$. For a circle, $h = r$, so $A = \frac{1}{2} \times 2\pi r \times r$. This is the same as πr^2."
Use dissection and area measure of triangle to generate, justify, and use area formula for regular polygons.	"If you split the hexagon into six equilateral triangles, and you know that the area of each triangle is $\frac{1}{2}bh$, where b = the length of the side of the hexagon, then the area of all six triangles is $\frac{1}{2}bh \times 6$, which is the same as $\frac{1}{2} \times perimeter \times h$ because $6b$ is the perimeter of the hexagon. This formula works for all regular polygons, if you think of each polygon as being made up of congruent triangles."
Use dissection and/or Cavalieri principle to generate and justify area formula for a triangle.	"The area formula for a triangle is $\frac{1}{2}(b \times h)$ because if you doubled the triangle and put the pieces together so that it formed a rectangle, the area formula would just be b × h. But since the original triangle is only half of the rectangle, it would be $\frac{1}{2}$ of the rectangle's area, or $\frac{1}{2}(b \times h)$."
Use dissection and/or Cavalieri principle to generate and justify area formula for a parallelogram.	"The area formula for a parallelogram is $b \times h$ because if you sweep the base through the height, you get $6 \times 4 = 24$ u². To check, if you took the triangle off one end of the parallelogram and put it on the other end so that it looked like a rectangle, then you would still have $6 \times 4 = 24$ square units for the area." "The parallelogram has the same base and the same height as a rectangle, so it is the same area." (Cavalieri principle)

resources that appear to be adequate to surmount these potential barriers. It is important both to root unit in acts of dissection that facilitate comparison of the magnitudes of areas and to avoid tasks that reduce area measure to covering and counting with preselected units. The objective is to create opportunity for considering properties of units that enable reliable and comparable measures of area.

Along the way, structuring the area as units or composites of units that are iterated to generate a measure of area creates opportunities to compare magnitudes of area to magnitudes of unit and composite unit or to compare magnitudes of the area of different enclosed regions of the plane. These comparisons can be both additive (how much more or less) and multiplicative (how many times more or less). The complementary, dynamic perspective revisits measure of area by affording opportunities to consider generation of area through motion of one length through another—a transformation in referent from one to two dimensions (Schwartz, 1996). This dynamic perspective supports developing conceptions of product as continuously generated, and composites of units are now enlivened as means for considering rates of production of magnitudes of area as a length is moved through another length. Cavalieri's principle provides a dynamic way to conceive of the rationale for commonplace area formulas. This conception supplements approaches that rely on dissection for justification. Dynamic generation and dissection are complementary perspectives that can be coordinated to help students come to see formulas as arguments, rather than as rote symbolic expressions.

References

Battista, M. T. (2004). Applying cognition-based assessment to elementary school students' development of understanding of area and volume measurement. *Mathematical Thinking and Learning, 6*(2), 185–204.

Battista, M. T., Clements, D. H., Arnoff, J., Battista, K., & Van Auken Borrow, C. (1998). Students' spatial structuring of 2D arrays of squares. *Journal for Research in Mathematics Education, 29*(5), 503–532. https://doi.org/10.2307/749731.

Brady, C., & Lehrer, R. (2021). Sweeping area across physical and virtual environments. *Digital Experience in Mathematics Education, 7,* 66–98. https://doi.org/10.1007/s40751-020-00076-2.

Cullen, A. L., & Barrett, J. E. (2020). Area measurement: Structuring with nonsquare units. *Mathematical Thinking and Learning, 22*(2), 85–115. https://doi.org/10.1080/10986065.2019.1608619.

Cullen, A. L., Eames, C. L., Cullen, C. J., Barrett, J. E., Sarama, J., Clements, D. H., & Van Dine, D. W. (2018). Effects of three interventions on children's spatial structuring and coordination of area units. *Journal for Research in Mathematics Education, 49*(5), 533–574.

Kobiela, M., & Lehrer, R. (2019). Supporting dynamic conceptions of area and its measure. *Mathematical Thinking and Learning, 21*(3), 178–206. https://doi.org/10.1080/10986065.2019.1576000.

Lakoff, G., & Núñez, R. E. (2000). *Where mathematics comes from: How the embodied mind brings mathematics into being.* New York, NY: Basic Books.

Lehrer, R. (2003). Developing understanding of measurement. In J. Kilpatrick, W. G. Martin, & D. E. Schifter (Eds.), *A research companion to principles and standards for school mathematics* (pp. 179–192). Reston, VA: National Council of Teachers of Mathematics.

Lehrer, R., Jacobson, C., Thoyre, G., Kemeny, V., Strom, D., Horvath, J. ... Koehler, M. (1998). Developing understanding of geometry and space in the primary grades. In R. Lehrer & D. Chazan (Eds.), *Designing learning environments for developing understanding of geometry and space* (pp. 169–200). Mahwah, NJ: Lawrence Erlbaum Associates.

Lehrer, R., Jaslow, L., & Curtis, C. (2003). Developing understanding of measurement in the elementary grades. In D. H. Clements & G. Bright (Eds.), *Learning and teaching measurement. 2003 Yearbook* (pp. 100–121). Reston, VA: National Council of Teachers of Mathematics.

Lehrer, R., Jenkins, M., & Osana, H. (1998). Longitudinal study of children's reasoning about space and geometry. In R. Lehrer & D. Chazan (Eds.), *Designing learning environments for developing understanding of geometry and space* (pp. 137–167). Mahwah, NJ: Lawrence Erlbaum Associates.

Lehrer, R., & Slovin, H. (2014). *Developing essential understanding of geometry and measurement for teaching mathematics in grades 3–5*. Reston, VA: National Council of Teachers of Mathematics.

Outhred, L. N., & Mitchelmore, M. C. (2000). Young children's intuitive understanding of rectangular area measurement. *Journal for Research in Mathematics Education*, *31*(2), 144–167.

Piaget, J., & Inhelder, B. (1969). *The psychology of the child*. New York, NY: Basic Books.

Piaget, J., Inhelder, B., & Szeminska, A. (1960/1981). *The child's conception of geometry*. New York, NY: W. W. Norton.

Schwartz, J. (1996). *Semantic aspects of quantity* (Unpublished manuscript). Cambridge, MA: MIT and Harvard Graduate School of Education.

Smith III, J. P., Males, L. M., & Gonulates, F. (2016). Conceptual limitations in curricular presentations of area measurement: One nation's challenges. *Mathematical Thinking and Learning*, *18*(4), 239–270.

Strom, D., Kemeny, V., Lehrer, R., & Forman, E. (2001). Visualizing the emergent structure of children's mathematical argument. *Cognitive Science*, *25*, 733–773.

Thompson, P. W. (2000). What is required to understand fractal dimension? *Mathematics Educator*, *10*(2), 33–35.

Thompson, P. W. (2011). Quantitative reasoning and mathematical modeling. In L. L. Hatfield, S. Chamberlain, & S. Belbase (Eds.), *New perspectives and directions for collaborative research in mathematics education. WISDOMe monographs* (Vol. 1, pp. 33–57). Laramie, WY: University of Wyoming. http://bit.ly/14w0flA

Zacharos, K. (2006). Prevailing educational practices of area measurement and students' failure. *Journal of Mathematical Behavior*, *25*(3), 224–239.

5

EXTENDING MOTION TO THREE DIMENSIONS

Volume and Its Measure

In typical elementary school mathematics curricula, learning about volume measurement is initiated by acquiring a formula for calculating the measure of right rectangular prisms: volume = length × width × height. Often, this approach fails to engage students' conceptual understanding of volume because it requires only that students recall and execute a simple arithmetic procedure (Simon & Blume, 1994). Evidence suggests that approaches that emphasize volume calculation are less successful than those that foster a conceptual understanding of solid volume measurement (Huang & Wu, 2019). A conceptual grasp of volume is important so that students can extend volume measure beyond the traditional focus on rectangular prisms to include prisms with other bases and cylinders and other settings of volume measure where the familiar formula no longer applies. Moreover, as in other realms of spatial measure, it is critical to help students develop a theory of measure organized around core concepts and anchored in practical activity.

Structuring and Dynamic Approaches to Volume Measure

Accordingly, mathematics education researchers have explored two general approaches to fostering students' intuitions and emerging conceptions of volume. The first, extensively pursued by Battista, Clements, and colleagues (Barrett, Clements, & Sarama, 2017; Battista, 1990, 1999, 2004, 2007, 2012; Battista & Clements, 1996, 1998; Clements, Swaminathan, Hannibal, & Sarama, 1999; O'Dell et al., 2017), supports students' structuring of rectangular prisms and compositions of prisms as a lattice of unit cubes. This entails first enacting and eventually imagining completely filling 3-D structures with cubic units of a standard size and enumerating the cubes to obtain a measure of volume.

DOI: 10.4324/9781003287476-5

Some (e.g., Panorkou, 2020) refer to this form of imagining as a packing or filling model, to reflect its emphasis on identifying the amount of substance that an object can hold. Battista and Clements (1996) describe the levels of understanding that children typically traverse as they participate in instruction designed to capitalize on this model. Beginners sometimes concentrate exclusively on tabulating the unit faces and fail to visualize them as parts of unit cubes that are three-dimensional. Or, students may try to enumerate cubes, but lack a systematic way of identifying and accounting for all the unit cubes because they do not have an integrated model of how the units are organized within the whole. As a result, they count some units twice and others (especially those within the interior of the prism) not at all. Eventually, however, students begin to identify and count composites of unit cubes and to organize these as layers that can be iterated to measure a volume. For example, the volume of a 3 u × 4 u × 4 u prism is conceived of as four iterations of the 3 u × 4 u × 1 u layer of 12 cubic units. In time and with appropriate instruction, students become more flexible at imagining these composites, first by partially structuring the prism with cubes needed to line the width, length, and height of the prism and eventually by visualizing the lattice structure, given only the dimensions of the prism. Once students can flexibly visualize all three dimensions from their measures, they begin to think multiplicatively—that is, to regard volume measure as a multiplicative relationship that links the three dimensions of the prism (and also of each unit cube).

Rather than emphasizing the accumulation of discrete units, the second approach instead emphasizes dynamic measure of 3-D structures by visualizing volume and units of volume as generated by sweeping, dragging, or pulling the area of the base of a 3-D structure continuously through its height. Lehrer, Strom, and Confrey (2002) described how a third-grade student proposed this image to members of his class. He described volume of a cylinder as "like pulling" the area of a base through its height. Subsequently, this student and his classmates relied on this image of pulling as they developed a way to find the volume of cylinders. Students accomplished this task by tracing the base of the cylinder on a square-unit grid and then identifying and counting square units of the base. Next, they estimated the remaining fractional parts of units, such as one-fourth of a unit, and then composed them into whole units (e.g., $\frac{1}{4}u^2 + \frac{1}{2}u^2 + \frac{1}{4}u^2 = 1u^2$). These recombined units were then added to the whole units that were initially counted to approximate the area of the base. Finally, students multiplied the estimated area of the base by the height of the cylinder "to draw it [that is, the area of the base] through how tall it is" (Lehrer, 2003, p. 186). Later, during one-on-one interviews, students were challenged to find the volume of an unfamiliar structure, a hexagonal prism, using a ruler and grid paper as tools. Most (75% of the 23 students) found the volume of the cylinder by first approximating the area of the hexagon with grid paper and then again drawing this area through the height of the prism to find the measure of its volume (Knapp & Lehrer, 2005).

A dynamic approach to volume measure helps children interpret a product volume as consisting of an arbitrarily large number of infinitesimally narrow layers or as continuous change, rather than only as discrete, countable units (Castillo-Garsow, Johnson, & Moore, 2013; Confrey, Maloney, & Corely, 2014; Lehrer & Slovin, 2014). Panorkou and colleagues (Panorkou, 2020; Panorkou & Pratt, 2016) amplify dynamic generation of volume with virtual environments designed to help students explore multiplicative coordination of quantities of area and height of prisms and cylinders (in this case, contextualized as candles). The goal of this work is to create an account of how students increasingly come to regard volumes (and units of volume measure) as multiplicative objects, defined by the quantities of area and height working together.

Thinking of volume measure dynamically has counterparts in length measure (see Chapter 3, this volume) and area measure (see Chapter 4), so dynamic images of continuity are conceptual resources that integrate otherwise disparate systems of spatial measure (Lehrer & Slovin, 2014). The learning trajectory for volume, which we next describe in detail, describes forms of student thinking that incorporate aspects of both the discrete and the continuous metaphors for conceiving of change in volume.

Students in our participating classrooms often begin explorations in area and volume by second or third grade, and explorations in area measure constitute important conceptual resources for investigations of volume measure. Developing conceptions of volume measure are supported by investigating properties of units, unit dissection of volumes, visualization of dynamic generation of volume as a product of area and length, and development of formulas guided by dynamic generation to find the volume measure of right and non-right prisms, as well as cylinders. In the earliest grades, instruction involves structuring volumes into units, using the kinds of filling metaphors and structuring tasks introduced by Battista and colleagues, with extension to dimensions that are not whole numbers to encourage students to begin to think of layers as becoming progressively "thinner" (e.g., where one dimension of a prism may be $3\frac{1}{2}$ u, then $3\frac{1}{4}$ u, …). This kind of thinking is a prelude to visualizing dynamic generation of volume and unit volume as a motion of area through length. Experiences with generating nets that describe three-dimensional figures support students' growing coordination of surface area and volume. Volume as dynamic sweeping has been successfully introduced as early as the third grade (Knapp & Lehrer, 2005), although instructional timelines governed by state standards often delay instruction on volume until fourth or fifth grade.

Volume Conceived as Space Inside

As students first consider volume, they informally describe it as space inside or contained by an object. Students differentiate surface area from volume, but their sense of both attributes is qualitative. Building on these informal conceptions of volume as space occupied, in the second level of the learning trajectory students begin to compare volumes of specific 3-D objects qualitatively, perhaps by nesting one within another or by filling containers with popcorn or sand and comparing the amounts needed to do so. For example, students in a second-grade class investigated folding an 8.5 × 11–inch sheet of paper into a container that could hold "the most popcorn." They were surprised to find that conical shapes held less popcorn than "boxes" (prisms). Table 5.1 summarizes these first two levels of conceptual development.

Measuring Volume by Accumulating Units

Shortly after these initial qualitative explorations, students begin to find and compare volumes by employing units. As with length and surface area, an important conceptual achievement in understanding volume is coming to understand the properties of units of volume. Students recognize that the units used to measure must be the same or, if not, distinctly labeled, and units need to tile, or completely fill, the volume being measured. Students' work at this level begins with rectangular prisms. Later, when they encounter other 3-D shapes (such as cylinders), the nature of units may need to be reconsidered, because students have a tendency to favor units that bear a perceptual resemblance to the figure being measured. For example, they may be drawn to using beads to fill

TABLE 5.1 Students Conceptualize Volume as Space Inside a Three-Dimensional Figure and Begin to Compare Volumes Qualitatively

Description	*Examples*
Compare volumes of different solids by filling them.	"The popcorn filled this first box, but the same amount of popcorn wasn't enough to fill the second box. The second box can hold more popcorn inside it than the first box can." "When we filled the first one with water, we used all of the water. When we filled the second one, some water was left over, so it doesn't hold as much as the first one."
Compare volumes of different solids using nesting.	"The red one fits inside of the blue one, so its volume is smaller."
Differentiate surface area from volume.	"The buildings have different numbers of windows (faces of cubes composing the structure), but it looks like they have about the same amount of space inside." "The outside of the cylinder is like a wrapped rectangle, but the cylinder has stuff inside."
Compare space inside or space contained by two or more objects.	"It looks like this one is bigger inside." "This one holds more than that one."
Recognize space inside or space contained by object.	"You can put something inside it."

the volume of a cylinder because both the beads and the cylinder are "curvy." However, students usually recall readily that they have already resolved this issue of resemblance in the context of area and quickly reject the solution based on similarity because it does not meet the criteria of avoiding both "gaps" and "overlaps" of measure so that units entirely fill the three-dimensional space.

As they first attempt to measure the volume of rectangular prisms, students find volume by counting unit cubes in structures in which all the cubes are visible. Much of the instruction in the early grades is aimed toward helping students differentiate the area of 3-D figures from their volume, but also to perceive the relations between these two attributes in any three-dimensional shape. In Volume unit 1 of Measuring and Visualizing Space, students compare the surface areas of three different "apartment buildings" (that is, prisms) constructed of interlocking cubes. Each building is composed of 12 cubic "apartments" with the following dimensions: $1 \times 1 \times 12$ cubes, $2 \times 2 \times 3$ cubes, and $6 \times 1 \times 2$ cubes. Surface area is contextualized as the windows, roof, and footprint of the building. Students next compare the volume enclosed or occupied by each building—that is, the number of "apartments" that each building holds. In this introductory task, the context supports children in differentiating surface area (e.g., the number of windows, roof, and building footprint) from volume (e.g., the total number of apartments enclosed in each building). Students also find that even though the surface areas of the three buildings vary, their volume measures are the same. The lesson was originally designed for classes of second-grade students, but it can also serve as an entrée to volume measure in later grades.

Subsequently, students develop strategies to visualize and account for cubes "hidden" within a more complex structure—for example, a building constructed from $4 \times 4 \times 4$ cubic-inch blocks, as shown in Figure 5.1. This task can be difficult at first if a child's strategy relies on counting units one by one, a strategy that can lead to confusion about what has and has not been counted. It is not unusual for students to count some cubes twice and overlook others altogether.

Counting cubes that are hidden within the structure and not directly perceptible is enabled by mentally imposing a lattice of unit cubes on the space

FIGURE 5.1 Accounting for cubes hidden within a structure

FIGURE 5.2 Student strategies for visualizing the structure of the volume as a lattice of cubes

contained by the structure. Students may begin by composing and recalling parts of the lattice (for example, enumerating the number of rows or columns), but in time and with experience they are able to visualize the entire rectangular prism as composed of or filled with units. As Figure 5.2 illustrates, students may structure the same volume in different ways, and the resultant strategies are discussed during subsequent classroom conversation. As with other measures, a volume measure is a ratio: a volume of 64 u^3 means that the volume is 64 times that of 1 u^3.

Visualizing Volume as Composites of Layers

As students learn to visualize and account for hidden cubes in a prism, they usually begin to visualize the rectangular prism as composed of layers, arranged either horizontally or vertically (as columns). The measure of the volume is obtained as the repeated addition of the number of cubes in each layer. Rather than needing to physically construct a complete unit dissection (e.g., constructing a 4 in × 4 in × 3 in prism with 48 cubic units), students can visualize the complete unit dissection with minimal support by configuring a portion of it. For example, students could represent the intersection of three edges of the 4 × 4 × 3 prism as in Figure 5.3 and then envision the complete lattice as three 4 in × 4 in horizontal layers or as four vertical layers, each of dimension 4 in × 1 in × 3 in, as shown in the figure. Students may also generate and use other composite units to determine the volume. Students compare the measure of the volume with the measure of a composite unit or unit; for example, 48 in^3 = 3 × 16 in^3. That is, the measure of the volume is three times as much as the measure of the volume of a 4 in × 4 in × 1 in layer.

The accomplishments just described encompass two levels in the learning trajectory for volume—the first addresses the development of students' understanding of volume units and their properties, and the second describes the strategies students acquire for partially structuring volumes. Both are described in Table 5.2.

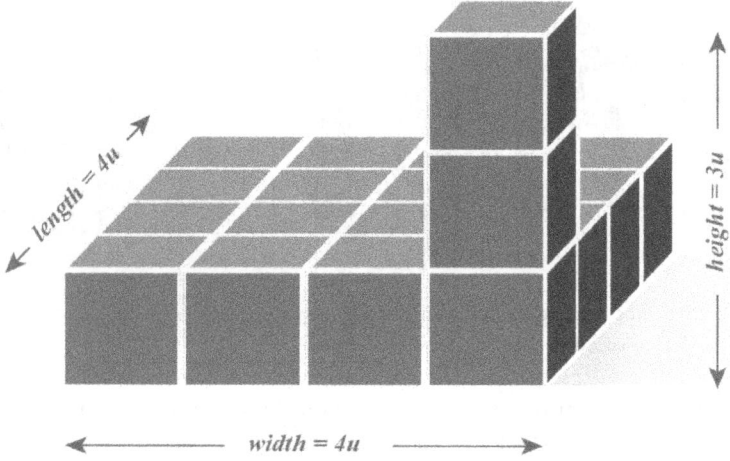

FIGURE 5.3 Finding volume by accumulating layers

TABLE 5.2 Students Understand Properties of Volume Units and Begin to Partially Structure Volumes

Description	Examples
Find and compare volumes of right rectangular prisms with whole-number dimensions using partial structuring strategies.	(For a 3 × 4 × 3 hollow prism without markings) "I can use the cubes to make a bottom layer of 12 and a height of 3, then multiply those numbers together to get 36 cubic units." "This box has a base of 12 and height of 3 = 36 cubic units, so it is smaller than the box that has a base of 10 and height of 4 = 40 cubic units."
Find and compare volumes of right rectangular prisms by counting unit cubes, including "hidden cubes."	(For a 3 × 3 × 4 prism with hidden unit) "I can count the cubes to find that there are 34 cubes on the outside, and then there would be 2 cubes you can't see on the inside, so the whole thing is 36 cubic units."
Create different right rectangular prisms of equal volume.	"These two take up the same amount of space, even though they do not look the same, because they each contain 36 cubic units."
Explain volume measure as a ratio.	"45 in³ means that the volume of this structure is 45 times that of the unit cube—the cube with length, width, and height of 1 inch, or 1 in."
Find and compare volumes of right rectangular prisms by counting unit cubes (no "hidden cubes").	(For a 2 × 2 × 9 lattice) "I can count the cubes to find that there are 36 cubes all together."
Recognize and explain properties of units (e.g., identity, tiling).	"Cubes are better than beans for filling the box, because they are all the same size with no gaps in between." "You have to use cubes because they take up the space inside. If you just drew squares to cover the box, the inside would still be empty."

To support visualization of unit dissection and composite unit dissection of prisms, students use manipulatives of unit cubes with each edge 1 in or 1 cm, at first to completely fill and subsequently to partially structure a prism. A second support for visualization of the lattice structure of prisms is to generate cubic units and compositions of cubic units by folding "nets" of prisms. In Volume unit 2, students are challenged to envision the surface area and volume of net representations of four different "apartment buildings." A net is a two-dimensional shape that can be folded into a three-dimensional figure. The teacher begins by demonstrating a net of a cubic unit, showing how the two-dimensional structure folds to create a three-dimensional cube. The teacher then challenges students to generate as many different nets as they can that will fold up to produce a cube of the same dimensions. The solution of one third-grade class is displayed in Figure 5.4.

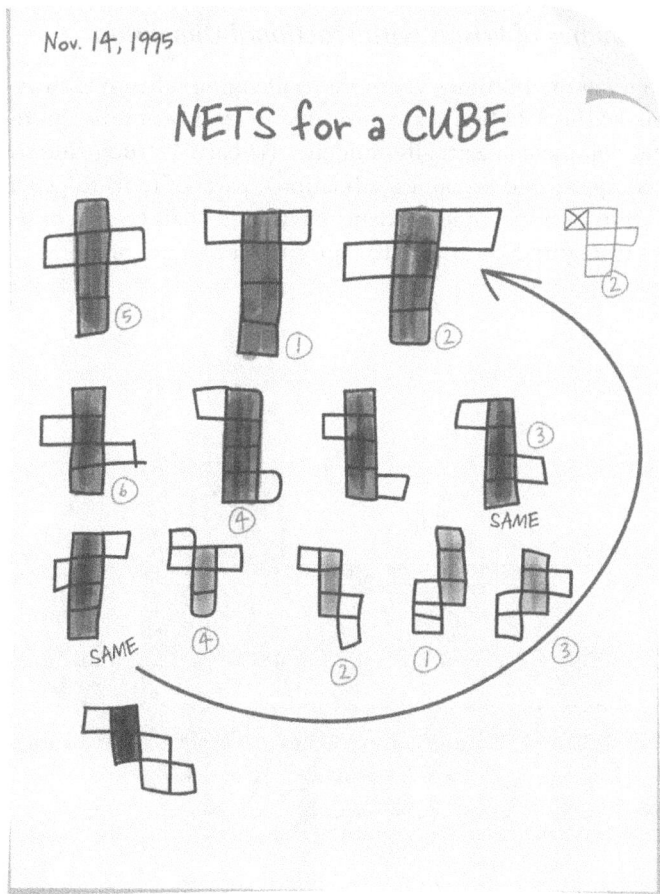

FIGURE 5.4   The proof generated by a third-grade class for 11 possible nets of a cube

The solution was made possible as the class articulated their criteria for deciding whether two different nets can be considered equivalent. Students usually decide, as these students did, that nets that can be made congruent via reflection, rotation, or translation can be considered "the same." The teacher displayed the nets created by pairs of students and asked the class whether there was a way to know whether they had discovered all the possible nets.

Children explained that they had generated all possible configurations by exhaustively searching columns with 4, 3, and 2 squares (which the students called backbones). Although this example concerns volume measure, it illustrates how measure can serve as a context for introducing other important ideas in mathematics, such as equivalence and explanations of why (that is, proof).

Finding Volumes of Prisms with Fractional Dimension

Visualizing volume becomes yet more challenging when one or more of the dimensions of the 3-D figure are measured in fractional units. In this case, to understand volume conceptually, students use partial structuring strategies to identify, compose, and account for fractional parts of units to generate a sum in whole units, For example, students envision a "half layer" consisting of 16 half-cubes in Figure 5.5a and of 12 half-cubes in Figure 5.5b.

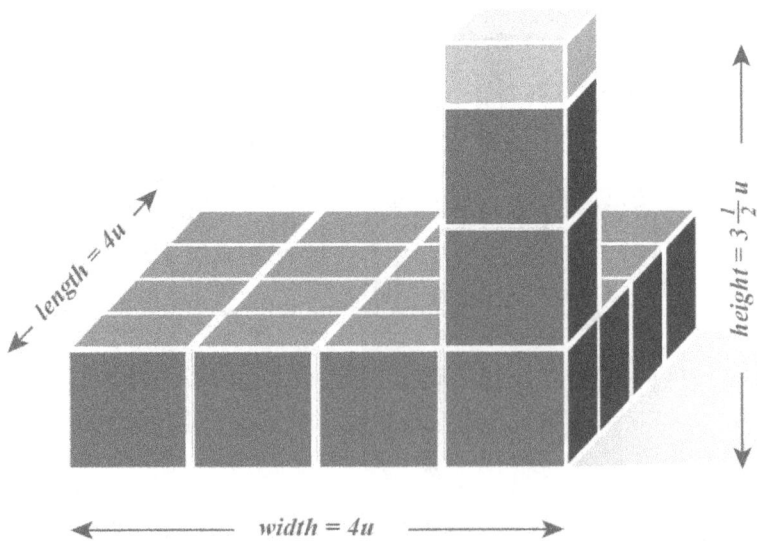

FIGURE 5.5a Rectangular prism with a height measured in fractional units

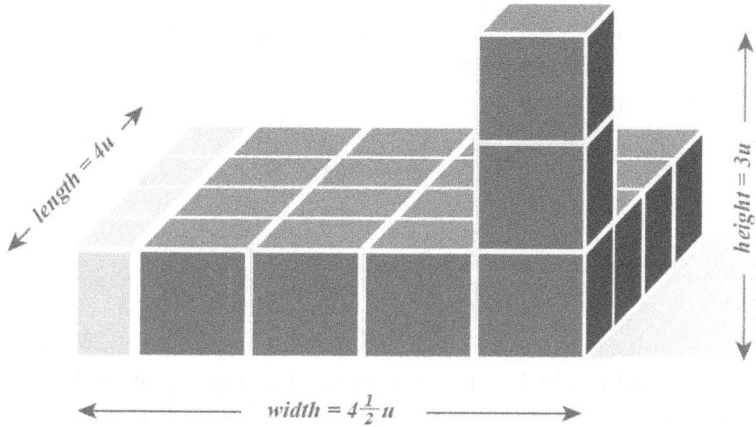

FIGURE 5.5b Rectangular prism with a width measured in fractional units

Table 5.3 describes and illustrates the conceptual advance that characterizes this level.

Generating Volume Dynamically

Students' visualization of volume shifts from an additive model, focused on identifying and accumulating layers, to a multiplicative model, in which both unit cubes and the volume of the entire prism are seen as composed of three interrelated dimensions: length, width, and height. Sweeping is used to envision the measure of the volume of a prism as a continuous product formed as the area is swept through the height, a perspective that entails thinking of volume as a continuous quantity. Students may conceive of a 3-D figure as a series of infinitesimally narrow layers of the base or, alternatively, may conceive of the base as being continuously stretched or pulled through its height. For example, the volume measure of a right rectangular prism with a base

TABLE 5.3 Students Find and Compare Volumes with Fractional Units

Description	Examples
Find and compare volumes of right rectangular prisms (including some with fractional dimensions) using units with partial structuring strategies.	(For a 3 × 4 × 4.5 hollow prism without markings) "I can use the cubes to make a bottom layer of 12 and a height of 4, that's 48, and then another layer that looks like half a cube (gestures cut cube), then add those to get 54 cubic units."

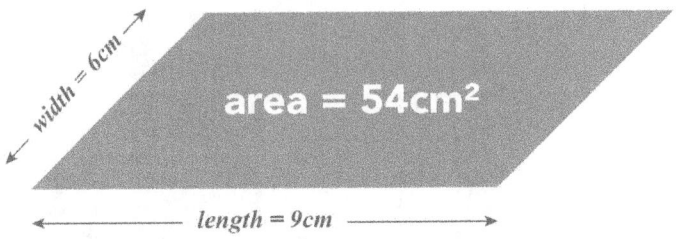

FIGURE 5.6a Students start with a 54 cm² base

of area 54 cm² and height 10 cm is conceived of as a sweep of the base area through the height, with the resulting volume of 540 cm³, as illustrated in Figures 5.6a and 5.6b.

As students revisit volume as a product, now regarded as generated continuously, teachers support this dynamic image of volume by encouraging students to parse common formulas by constructing drawings of their interpretation of the volume measure as the area of the base of a prism moving through the height. For example, during a formative assessment conversation, a teacher asked a fifth-grade student, Cas, to draw what length × width meant for a 4 in × 6 in prism with height 10 in. Cas used a ruler to draw a 4 in × 6 in rectangle and mentioned that the measure of its area was 24 in².

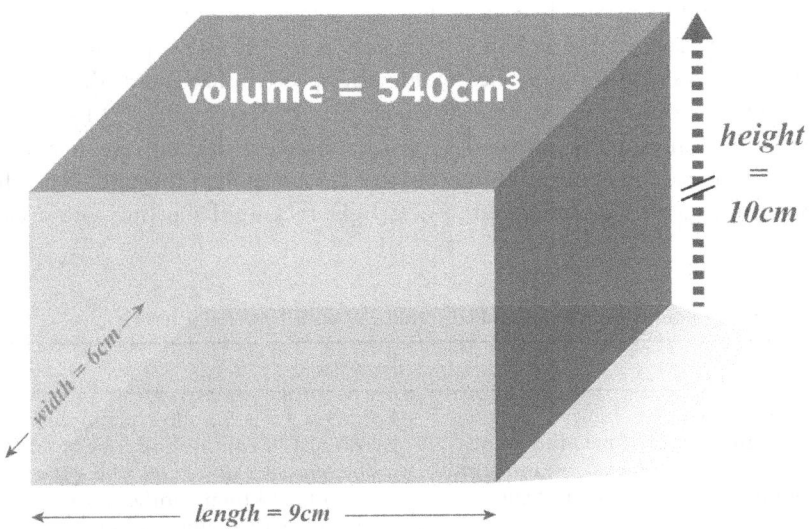

FIGURE 5.6b Base is swept through the height of 10 cm, resulting in a measure of 540 cm³

She demonstrated the rectangular base's unit dissection by coordinating the intervals of length measure on each side (a performance that is consistent with the third level of the Area construct) to create 24 square units. Then the teacher asked Cas to draw 4 in × 6 in × 1 in, and in response Cas drew a representation of 24 cubic units. The teacher continued to encourage Cas to add increments of 1 in in height.

To further emphasize the continuous production of volume, the teacher asked the class to imagine dragging the base "up through a height of one-half inch." "What," she asked, "would then be the measure of the volume?" Some students claimed that the volume would be 12 in^3 because the first layer, the composite of 24 cubic units, was now sliced in half. Others argued that it was better to think of the 12 cubic units as growing from the square units in the base to a height of one-half inch, so that the 12 cubic units were the result of combining 12 half-cubic units. Students became excited by this prospect and went on to investigate the volume measure of movement of the base through ever-decreasing intervals of height, each interval one-half times as long as the previous height (e.g., $\frac{1}{2}$ in, $\frac{1}{4}$ in, $\frac{1}{8}$ in, $\frac{1}{16}$ in, …). Some students pursued this process through $\frac{1}{2048}$ in. Engaging in this discussion helped these students further elaborate the meaning of continuous change in terms of what convention would call an infinitesimal amount, an arbitrarily thin slice (or, as one of the fifth-grade investigators said, "as thin as you like").

To further consider the multiplicative coordination of dimension implied by movement (or sweeping, as students called it), students were challenged to find ways of doubling the volume of the prism. Most proposed doubling a single dimension, and during a class conversation one of the students gestured to indicate how doubling a dimension of the prism would result in two copies of its volume. One student's proposal to double the length of every dimension led to a new multiplicative relation, where the volume of the "stretched" prism was 8 times the volume of the original.

As their experience with sweeping area through height becomes elaborated, students find and compare volumes of other right prisms, such as those with triangular and hexagonal bases, as well as cylinders and composite structures. For example, the volume of a triangular prism can be thought of as the product of the area of the triangular base and the height of the prism.

Finally, students use Cavalieri's principle to explain why the volume measure of non-right prisms and cylinders is equal to the measure of their corresponding right prisms and cylinders. Examples of student reasoning at this level are featured in Table 5.4.

Students' reasoning about oblique prisms often builds from their visualization of prisms as a series of very thin slices that result when the base of a prism is drawn through its height. Students frequently argue that the volume of an oblique prism is the same as the volume of its corresponding right prism

TABLE 5.4 Students Find and Compare Volumes Dynamically

Description	Examples
Explain and use Cavalieri's principle to find the volume of different non-right prisms and cylinders.	(For a 3 × 4 × 4.5 hollow prism without markings) "I can use the cubes to make a bottom layer of 12 and a height of 4 that looks like half a cube, then multiply those to get 54 cubic units."
Find and compare volume of a variety of prisms.	"If the area of the base is about $3\frac{1}{2}$ square inches and the height is 4 inches, then I would say that the volume of the hexagonal prism is base area × height, or 3.5 times 4, which is about 14 cubic units."
Find and compare volumes of right cylinders using sweeping of area through height.	"If the area of the base is about 3 squares and I move that through the height of the side, 11 cubes, then I would get about 33 cubic units."
Generate, use, and explain formulas for volume of right cylinders. Find and compare surface areas of right cylinders using sweeping.	"So if the area of the base is about 3 squares and the height is 11, then you can just say that the base area times the height equals the volume, so 3 × 11 = 33 cubic units." "If I wrap the grid paper around the cylinder, the distance around the base [circumference] is about 9 units, and then if I pull that through the height of 16, I get 9 × 16 = about 144 square units for the surface area of the side of the cylinder."
Find and compare volumes of right rectangular prisms (including some with fractional dimensions) using sweeping of area through height.	(For a 3 × 4 × 4 hollow prism without markings) "If the area of the base is 12 squares, and I move that all the way through the height of the box, 4, then I would get 48 cubic units."
Generate, use, and explain formulas for volume of right rectangular prisms.	"So if the area of the base is 12 and the height is 4, then you can just say that the base area times the height equals the volume, so 12 × 4 = 48 cubic units."
Justify sweeping as a strategy for finding volume.	"If the area of the base is length × width, or 3 × 4, and I layer that 4 times for the height, I would get 48 cubic units, or 3 × 4 × 4 = 48."

because very thin slices or layers of volume translated parallel to the base do not change their volume under this transformation. Students have justified this argument by referring to a stack of playing cards that has a constant volume, demonstrating that the cards can take on different configurations with the same base, as shown in Figure 5.7.

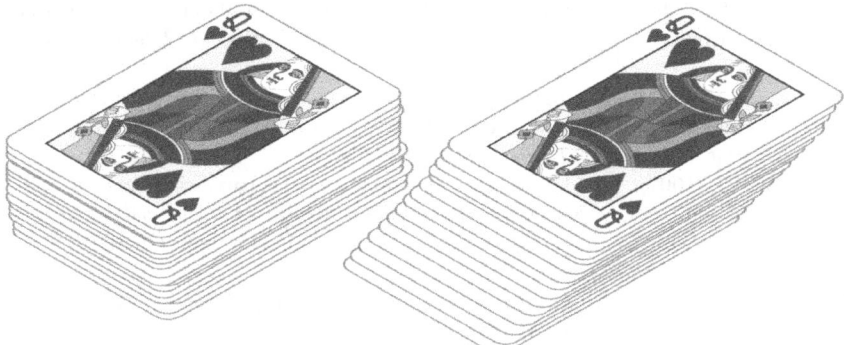

FIGURE 5.7 A stack of playing cards configured differently nonetheless has the same base and height and, therefore, the same volume

Students also begin to reason about structures that are not rectilinear, such as the volumes of cylinders, which can also be imagined as created by sweeping the base through the height. Students are sometimes surprised to discover that "unwrapping" the cylinder produces a rectangle that corresponds to the body of the cylinder (with dimensions equal to the circumference of the cylinder and its height). The area of this rectangle, plus the area of the two bases, constitutes the total surface area of the cylinder.

As the preceding examples illustrate, volume measurement both builds from and is consistently linked to understanding of area measure. The same principles of units (e.g., consistent units, tiling) are germane to measuring both area and volume. The dynamic metaphor of an area pulled through its base repeatedly reinforces both distinctions and connections between area and volume during instruction. There is good evidence that even much older students struggle to differentiate surface area from volume of polygons (e.g., Chiphambo & Mtsi, 2021), suggesting the value of (1) framing contexts that ground this differentiation to concrete meaning and (2) repeatedly articulating the conceptual and mathematical relationships between area and volume.

References

Barrett, J., Clements, D. H., & Sarama, J. (2017). *Children's measurement: A longitudinal study of children's knowledge and learning of length, area, and volume.* Journal for Research in Mathematics Education Monograph Series (Vol. 16). Reston, VA: National Council of Teachers of Mathematics.

Battista, M. T. (1990). Understanding students' thinking about area and volume measurement. In D. H. Clements & G. Bright (Eds.), *Learning and teaching measurement: 2003 yearbook* (pp. 122–142). Reston, VA: National Council of Teachers of Mathematics.

Battista, M. T. (1999). Fifth graders' enumeration of cubes in 3D arrays: Conceptual progress in an inquiry classroom. *Journal for Research in Mathematics Education, 30*(4), 417–446.

Battista, M. T. (2004). Applying cognition-based assessment to elementary school students' development of understanding of area and volume measurement. *Mathematical Thinking and Learning, 6*(2), 185–204.

Battista, M. T. (2007). The development of geometric and spatial thinking. In F. K. Lester, Jr. (Ed.), *Second handbook of research on mathematics thinking and learning: A project of National Council of Teachers of Mathematics* (pp. 843–908). Charlotte, NC: Information Age Publishing.

Battista, M. T. (2012). *Cognition-based assessment and teaching of geometric measurement. Building on students' reasoning.* Portsmouth, NH: Heinemann.

Battista, M. T., & Clements, D. H. (1996). Students' understanding of three-dimensional rectangular arrays of cubes. *Journal for Research in Mathematics Education, 27*(3), 258–292.

Battista, M. T., & Clements, D. H. (1998). Students' understanding of three-dimensional cube arrays: Findings from a research and curriculum development project. In R. Lehrer & D. Chazan (Eds.), *Designing learning environments for developing understanding of geometry and space* (pp. 227–248). Mahwah, NJ: Erlbaum.

Castillo-Garsow, C., Johnson, H. L., & Moore, K. C. (2013). Chunky and smooth images of change. *For the Learning of Mathematics, 33*(3), 31–37.

Chiphambo, S. M., & Mtsi, N. (2021). Exploring grade 8 students' errors when learning about the surface area of prisms. *EURASIA Journal of Mathematics, Science, and Technology Education, 17*(8), em1985.

Clements, D. H., Swaminathan, S., Hannibal, M. A. Z., & Sarama, J. (1999). Young children's concepts of shape. *Journal for Research in Mathematics Education, 30*(2), 192–212.

Confrey, J., Maloney, A. P., & Corely, D. (2014). Learning trajectories: A framework for connecting standards with curriculum. *ZDM: The International Journal on Mathematics Education, 46*(5), 719–733.

Huang, H.-M. E., & Wu, H.-Y. (2019). Supporting children's understanding of volume measurement and ability to solve volume problems: Teaching and learning. *EURASIA Journal of Mathematics, Science and Technology Education, 15*(12), em1789. https://doi.org/10.29333/ejmste/109531.

Knapp, N., & Lehrer, R. (2005, June). Changes in children's conceptions of spatial measure: Coordinating talk and inscription. Paper presented in *Understanding, building, and using symbolic representations of space and time* (M. Wiser, Organizer) at the 35th annual meeting of the Jean Piaget Society, Vancouver, Canada.

Lehrer, R. (2003). Developing understanding of measurement. In J. Kilpatrick, W. G. Martin, & D. E. Schifter (Eds.), *A research companion to principles and standards for school mathematics* (pp. 179–192). Reston, VA: National Council of Teachers of Mathematics.

Lehrer, R., & Slovin, H. (2014). *Developing essential understanding of geometry and measurement in grades 3–5.* Reston, VA: National Council of Teachers of Mathematics.

Lehrer, R., Strom, D., & Confrey, J. (2002). Grounding metaphors and inscriptional resonance: Children's emerging understanding of mathematical similarity. *Cognition and Instruction, 20*(3), 359–398.

O'Dell, J. R., Barrett, J. E., Cullen, C. J., Rupnow, T. J., Clements, D. H., Sarama, J. … Beck, P. S. (2017). Using a virtual manipulative environment to support students' organizational structuring of volume units. In E. Galindo & J. Newton (Eds.), *Proceedings of the 39th annual meeting of the North American chapter of the International Group for the Psychology of Mathematics Education* (pp. 1329–1336). Indianapolis, IN: Hoosier Association of Mathematics Teacher Educators.

Panorkou, N. (2020). Dynamic measurement reasoning for area and volume. *For the Learning of Mathematics, 40*(3), 9–13.

Panorkou, N., & Pratt, D. (2016). Using Google SketchUp to research students' experiences of dimension in geometry. *Digital Experiences in Mathematics Education, 2*(3), 199–227.

Simon, C. P., & Blume, L. (1994). *Mathematics for economists*. New York, NY: Norton.

6

INTEGRATING FIGURE AND MOTION IN THE MEASURE OF ANGLE

Dynamic and Figural Perspectives

A robust concept of angle integrates and coordinates two overarching perspectives (Henderson & Taimina, 2001) that are likely, especially with youngsters, to be distinct cognitive models (Clements, Battista, Sarama, & Swaminathan, 1997; Devichi & Munier, 2013; Keiser, 2004; Lehrer, Jenkins, & Osana, 1998; Masuda, 2009). The first, sometimes referred to as the *dynamic perspective*, interprets angles as rotation around a vertex. This perspective is dynamic in that the defining feature of the mental representation is rotation or turn. Rotation is constrained by a starting position and an ending position, but in the dynamic model, the turn is more salient than its boundaries. The turn of a doorknob is an example, as is the amount of rotation accomplished when an individual stands at a fixed point and then turns to face a new direction. In contrast, the second perspective, the *figural* or *static perspective*, represents angles as space delineated by intersecting lines. Examples include the diagrams of angles that appear in textbooks or the bend of a road represented on a map. In this perspective, the sides of angles are explicitly represented, and turn is less obvious. Perhaps because angles-as-figures are the prototype most often seen in elementary school mathematics texts, the rotational aspect of angles tends to be underemphasized in instruction. Accordingly, many researchers (e.g., Bryant, 2009; Kaur, 2020; Mitchelmore, 1998) find that young students typically fail to conceptualize turning in terms of angles. Instead, "turning" is thought of as an altogether separate category of phenomena.

There is general agreement that a goal of instruction about angle is to help students develop a rich background of models that can be applied to a variable range of situations, but there is less agreement about the best way to

DOI: 10.4324/9781003287476-6

accomplish this objective. Mitchelmore and White (2000) advocate starting with cases that students find easiest to perceive. They describe these cases as static situations and representations in which both sides of the angle are perceptually apparent. Over time, students are then gradually exposed to angle in less familiar settings, including those that involve rotation, and helped to recognize similarities among the exemplary cases. As children's angle concept grows by accretion and generalization, they gradually learn to connect those contexts. As more cases are experienced and analyzed, existing mathematical concepts begin to be embedded in explicit definitions. In contrast, other researchers advocate exposing children from the beginning to a rich range of contexts and kinds of angles and explicitly discussing how they are similar (e.g., Keiser, 2004). Still others argue for starting with general mathematical definitions and then subsequently showing students how concrete cases meet the definitional constraints. However, as Keiser (op. cit.) has pointed out, defining angle in a sufficiently general way is not easy, because the facets of angle that are central to a definition often vary with the context.

To surmount these complexities, researchers have explored the potential of dynamic geometry environments, including programming languages like Logo (Papert, 1980), to help children draw connections between dynamic and figural examples of angle (Crompton, 2015; Kaur, 2013, 2020; Latsi & Kynigos, 2011). As the Logo turtle moves, it leaves behind a trace that records its path. When the turtle turns from an ongoing straight path, the resulting trace creates a relation between two lines, one line indicating the previous heading and the other implied by the new heading. When the turtle moves again in the direction of the new heading, a trace is produced that is simultaneously both a turn (the turtle has just turned) and a space delineated by intersecting lines, a figure. However, the figure is not that of the turtle's turn from one heading to another, but is instead the path traveled by the turtle. Because Logo turns are expressed in parameters, students directly address questions about how to quantify, or measure, "amount of turn" as they write programs to control the turtle's path. The simultaneous motion and trace of the turtle implicitly link the figural and rotational aspects of angle, but one ray of the turtle turn angle is not represented explicitly and hence must be imagined for the motion to be perceived as a figure.

Research with elementary school students suggests that early optimism about Logo as a tool for angle instruction does not automatically lead to an integration of angle-as-rotation and angle-as-turn. For instance, as fourth graders pursued self-directed projects with Logo, they struggled to relate the Logo turn parameter to the angle formed by the turtle's path (Clements & Battista, 1990). Even extended periods of activity with Logo had limited effect on the development of many students' angle concepts (Clements & Battista, op. cit.; Mitchelmore, 1998). Clements and Battista (op. cit.) reported that 40 sessions with Logo did little to improve fourth-grade children's performance on the kinds of angle identification items that appear on standardized tests.

Another proposed way for helping to link motion to figure is to emphasize students' own bodily experiences in moving through space, so that they are made more directly aware of the relationship between static representations of angles and physical rotation. For instance, Clements and Burns (2000) added enactment with bodily turns to students' instruction with turtle turns in Logo. Students were encouraged to rotate their own bodies to build a physical intuition of turning, including both direction of turn (left or right) and amount of turn (expressed as a full turn or a fraction of a turn). They then used these benchmark turns to assign values to static representations of angles. In both static and dynamic contexts, students synthesized body movements and numerical estimates to judge turn measures, used part-turns like a half turn or quarter turn as benchmarks, and estimated the measure of static angles through guess-and-check strategies. As students brought these ideas into Logo programs, Clements and Burns (op. cit.) noted that they eventually began to replace full rotations of their bodies with increasingly abbreviated rotations of a hand or even a finger. The researchers proposed the term "curtailment" to refer to the gradual construction and manipulation of these part-gestures, eventually replaced by mental images, to refer to what had originally been a fully embodied enactment.

Similarly, Wilson and Adams (1992) proposed a sequence in which students first practiced making full, half, and quarter turns. Subsequently, they were encouraged to map those fractional measures of angles onto angles in a wide range of real-world contexts, including limited and unlimited rotation, meeting, inclination, corner, turning, direction, and opening. Although researchers reported some successes with this embodiment approach to linking turns and figural angles, they also acknowledged that the strategy was not a panacea. As Clements and Burns (op. cit.) point out, perceiving angle in the turns of one's own body places a nontrivial cognitive load on children's imaginative and representational capacities. To use bodily rotation to conceptualize an angle, children must construct and then maintain a memory of both the initial heading and the final heading, using a frame of reference in the context to fix these headings. At the same time, they need to maintain a record of the activity of rotation from the initial to the final heading and, finally, compare that rotation to internalized benchmarks of a turn (e.g., familiar turns like a quarter or a half turn). It is not surprising that children often struggle to master and integrate these operations. Providing opportunities for children to enact polygon paths as bodily movements and to re-represent these bodily movements with forms of Logo that annotated turns to make visible their figural aspect appeared to help third- and fourth-grade students view angles in a more integrated way (Lehrer, Guckenberg, & Lee, 1988; Lehrer, Randle, & Sancilio, 1989). Upon initiation of a student's command to turn the turtle, the turtle first sprouted a dashed line to annotate explicitly its pre-turn heading. This was followed by a slow-motion turn of the turtle to the new heading specified by the turn command. For example, input of the command TR $\frac{1}{4}$ or TR 90 initiated construction of an imagined continuation of the path as a dashed line at the current heading of the turtle, followed by a rotation of 90° clockwise at the turtle's current position. Related elaborations of Logo enabled students to use turtle turning at the vertex of static angles to obtain angle measure in degrees. These forms of intervention increased student facility at estimating the angle measure of angles-as-figures (Lehrer et al.,1989).

Early in the development phase of our project, we collaborated with partici-pating teachers to experiment with some of these prior approaches to angle that involved bodily rotation around a fixed point. We observed that the embod-ied-turn approach suffered from the challenges that we have just described, and as we explain in further detail in Chapter 8, these strategies seemed to impose a cognitive load that was daunting for first- and second-grade students. Lehrer and Pritchard (2002) suggested that involving children in mapping and wayfinding with their own bodies allows them to experience space, along with movements of moving and turning, more immediately than they do with Logo, which requires children to project themselves into the turtle's perspective. We there-fore designed tasks in which students walk and represent a variety of geometric paths (e.g., a rectangle, an equilateral triangle, a regular hexagon), so that angle is introduced in the context of geometry. We expected that doing so would capitalize both on Logo's emphasis on notating dynamic movement to construct length and angle and on the immediacy of embodiment. The geometric figures that children produce by walking a path that inscribes the figures' perimeters serve both as a mental prototype that guides their descriptions of pacing and angle and as a trace that can be interrogated to evaluate the adequacy of that description (e.g., When someone else walked according to my directions, did that person indeed produce a rectangle?). Two coordinated forms of notation are pursued in instruction (see Angle units), including ribbons and cones to record the path that is being walked and to differentiate bodily headings before and after turning, and written directions that specify direction and distance that the "walker" travels. In later elementary grades, embodied and imagined paths of polygons ground theorems about angles and, extended to wayfinding, provide students with opportunities to experience angle in the context of new notational systems, such as mapping and polar coordinates, that describe motion within a broader expanse of space. The resulting angle trajectory capitalizes on the most promising features of earlier approaches, which emphasize the value of embodiment and notational support for dynamic movement, while simulta-neously situating angle instruction within the broader mathematical frame of geometry. We anticipate that as robots and coding become more commonplace in schools, the fusion of bodily movement and dynamic inscription previously afforded by Logo may reinvigorate support for developing robust conceptions of angle, albeit again with explicit support for inscribing the paths produced by the code to highlight integration between dynamic and figural conceptions of angle. In the next section, we describe a trajectory of conceptual development that is supported when students are provided opportunities to consider angles-as-figures and angles-as-rotations in classrooms designed to support integration of these distinctive perspectives.

Benchmarks in Thinking About Angle

The trajectory for conceptions of angle consists of five distinct benchmark levels of conceptual development, along with related supports for learning. At each level, students explore and further coordinate the two related models for angle, angle-as-figure and angle-as-turn. The trajectory begins with examples of angles that young children notice in their everyday worlds, although at the beginning children are unlikely to understand these cases as members of a common category.

Noticing Canonical Examples

Through the primary grades, children recognize and refer to perceptual attributes of figures and structures that a knowledgeable adult would regard as examples of angles. However, for children, these noticings are images of a series of independently experienced clusters of cases that are not organized into a coherent, overarching conceptual system. Nonetheless, they serve as conceptual resources that provide starting points for elaboration as concepts of angle. For instance, children identify "corners" of structures (e.g., the right angles of intersecting walls or of buildings, the intersecting lines of floor tiles or of figures like squares on a printed page), or "pointy parts" (e.g., the vertices) of two- and three-dimensional shapes. They describe paths with angles as "bent" or "crooked." Straight paths that are not oriented horizontally or vertically are often considered not-straight (Olson, 1996). These images, which they usually do not yet regard as related, are seeds of the thinking that can later lead to interpreting angles as figures. Children also recognize and describe common experiences of rotation, such as "doing a 180" on a skateboard or twisting their bodies. These experiences of bodily motion are resources that can eventually lead to interpreting angles as turns.

Representing Angles-as-Figures, Angles-as-Turns

Early formal instruction tends to portray angles as they are commonly represented in diagrams and drawings. These symbolizations are starting points for beginning to consider angles as mathematical ideas, rather than solely as collections of aspects of things in the world, such as corners. Engaging students in creating representations of angle-as-figure and angle-as-turn initiates mathematization of what has previously been understood intuitively. Table 6.1 describes some of the

TABLE 6.1 Emerging Recognition of Angle-as-Figure and Angle-as-Turn

Performances	Examples
Classify angles as acute, right, obtuse.	"Obtuse angles are more than 90°, right angles are 90°, acute angles are less than 90°." Visually recognizes angles that are acute, obtuse, and right (e.g., given a set of angles, labels them correctly).
Represent angles as the intersection of line segments and compare angles by extent of openings (this may lead to either accurate or inaccurate judgments).	
View angle as bodily turn or as turn of a figure or solid. Measure turn as directed rotation from initial heading—for example, right (CW) or left (CCW)—in parts of a whole turn (TR $\frac{1}{2}$ whole turn). Represent magnitude and direction of turn with directed arcs or related symbols.	"I turned 1 whole turn to my left."  "I turned $\frac{1}{2}$ whole turn to my right." "I can turn my body 1 whole turn, like this." "I turned $\frac{1}{2}$ whole turn, which is less than 1 whole turn." "I turned 1 whole turn to my left." "TR $\frac{1}{2}$ wt means turn right one-half of a whole turn."

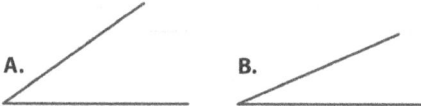

FIGURE 6.1 Comparing angles by perceiving the extent of opening between the endpoints of line segments

early conceptual seeds of children's emerging recognition of angles-as-turns and angles-as-figures.

Angle-as-figure is canonically represented as two intersecting lines. Children tend to compare angles-as-figures by the relative extents of the "openings" between the endpoints of the intersecting line segments of the figure (Lehrer, Jenkins, & Osana, 1998). For example, in Figure 6.1, angle A is considered greater than angle B because it is more open. Judging "openness" is a reliable way of distinguishing between angles and conforms well with intuitions about amounts of space formed by intersecting lines, but only if the lengths of the line segments that compose each angle are the same or nearly so.

However, focusing on the relative opening to compare angles-as-figures can lead to misjudgments. For example, in Figure 6.2a, angles C and D have the same degree measure, but children often consider angle C to be "bigger" (greater) than angle D, perhaps because the entire figure is larger. Similarly, in Figure 6.2b, angle E may be judged to be greater than angle F, although once again, each has the same measure in degrees.

Children may also be influenced by the orientation of angles-as-figures, believing that different orientations of intersecting lines imply different angles. Students at this second benchmark level are often familiar with correspondences between extents of openings and labels for angles as *acute*, *right*, and *obtuse*. They may even associate these figures with degree measures, such as *less than*, *equal to*, or *greater than 90°*, respectively. However, it is unusual for students at this level to offer a definition of a degree; they are likely instead to treat *degree* merely as a label that differentiates, for

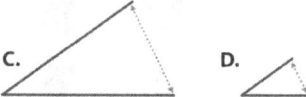

FIGURE 6.2a Comparisons between angles may be influenced by their apparent size

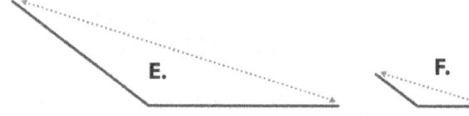

FIGURE 6.2b The comparative length of the openings suggests that E > F

example, corners from other kinds of intersections. An alternative image of angle—that is, angle-as-rotation—is conveyed by bodily turns, as in Figures 6.3a–d, with direction (to the person's right, or clockwise, contrasted to the person's left, or counterclockwise) and magnitudes of rotation (typically one whole turn versus $\frac{1}{2}$ whole turn) distinguished. These bodily experiences can be represented with arcs, as depicted in Figure 6.3, and these representations reconfigure ordinary experiences of turning into angles-as-turns.

Engaging children in inscribing bodily experiences of turning differentiates direction from magnitude of turn and introduces children to quantifying turn in relation to one complete turn. For example, TR 1 whole turn specifies a clockwise direction about an imagined vertical axis of one's body with a measure of one complete bodily rotation—the heading or direction of gaze is

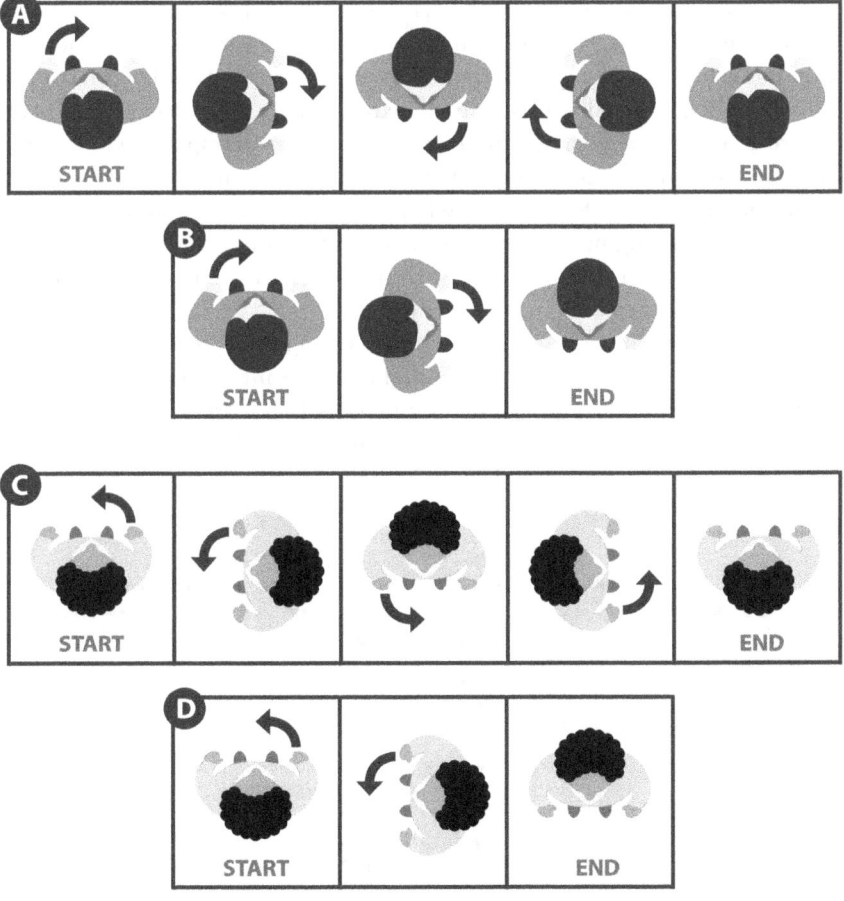

FIGURE 6.3   (a–d) Top view of bodily turns

the same before and after the turn. However, as yet, children do not usually have a firm grasp on other splits of turns, such as $\frac{1}{4}$ whole turn. As a result, all fractions of turns may be characterized as $\frac{1}{2}$ whole turn. Rather than using fractions to indicate measure of turn, some children will be familiar with landmark measures, such as the fact that a quarter turn has a measure of 90° and a half turn has a measure of 180°. However, as noted earlier, children usually understand these degree measures as categories, not as an iterated measure in degrees.

Integrating Angle-as-Turn with Angle-as-Figure, Interior vs. Turn Angles

At Level 3 of the Angle construct, students integrate the representations and measure of angle-as-form and angle-as-figure, a coordination that develops incrementally as students walk and symbolize paths describing familiar polygons, such as squares, rectangles, and equilateral triangles. Often in coordination with their increasing understandings of splits of units of length, students develop 2-splits of bodily turns, and they relate these splits to degree measure, where 1° is interpreted as $\frac{1}{360}$ th of a body turn. For example, $\frac{1}{4}$ whole turn is the same amount of rotation as 90° (90°/360°) and is understood as 90 iterations of a turn of 1°. Similarly, $\frac{1}{2}$ whole turn is the same amount of rotation as 180°. Comparing measured magnitudes of rotation, a 360° rotation is four times as much as a 90° rotation. Students relate turn angles to "bends" in paths, like those depicted in Figure 6.4.

Integrating turn and figure conceptions of angles is promoted by conceptualizing the measure of an angle-as-figure as the measure of the amount of turning needed to bring one line segment of the angle-as-figure onto the other line segment of the angle, as shown in Figure 6.5.

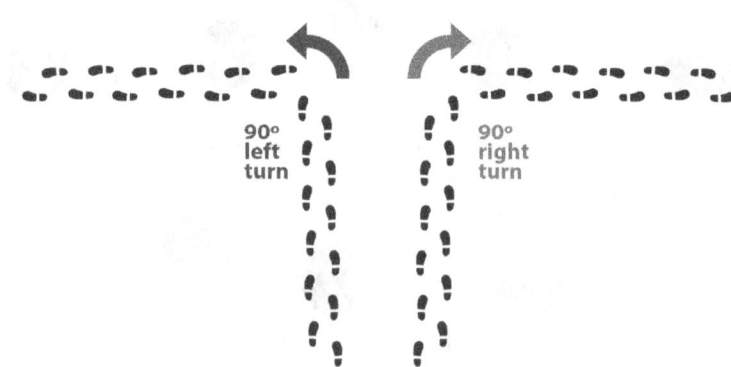

90°
left
turn

90°
right
turn

FIGURE 6.4 A bend in a path is generated by a turn

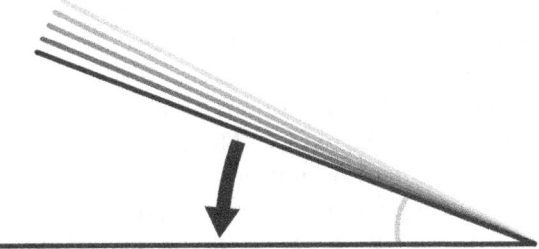

FIGURE 6.5 The measure of an angle-as-figure is the amount of rotation needed to bring one segment onto the other

This integration of figure and turn follows from the metaphor of polygons as paths, as introduced in Angle unit 1 of the instructional materials (https://www.routledge.com/Measuring-and-Visualizing-Space-in-Elementary-Mathematics-Learning/Lehrer-Schauble/p/book/9781032262734). If one walks a polygon path, the external angle of the polygon is the turn angle at a vertex. The interior angle is formed by the intersection of the sides at that vertex. Figure 6.6 depicts an equilateral triangle as a walking path. The exterior angle is the turn angle at each vertex. The turn angle can be thought of as an angle-as-figure if one imagines one side of the angle as the straight segment that

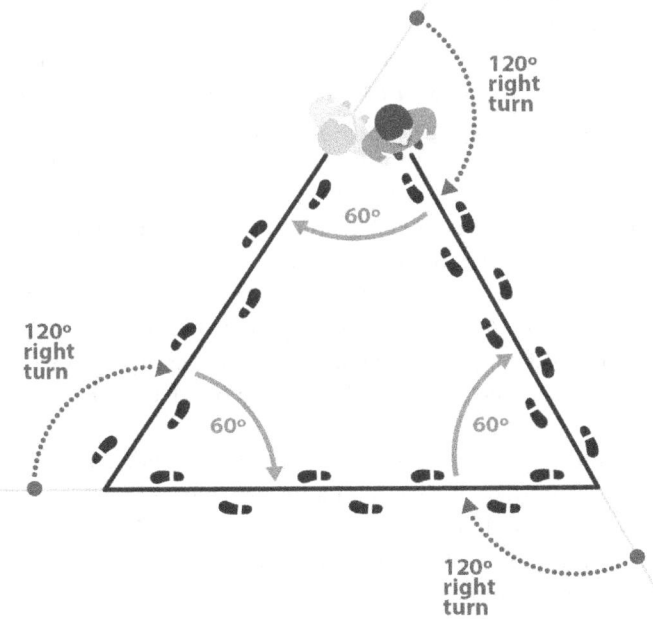

FIGURE 6.6 Equilateral triangle as a path with a 120° turn at each vertex

would result if the walker kept walking and did not turn (as indicated by the light grey line extension in Figure 6.6). The second side of the angle is formed by the side of the triangle that is walked after the turn. The interior angle is formed at each vertex by the intersection of the two line segments defined by the corresponding sides of the triangle. Notice that because the interior angle and exterior angle together constitute a straight line with a measure of one-half turn, or 180°, each interior angle of an equilateral triangle must be 60°. The supplementary relation between turn and interior angle follows from a symmetry of straight: straight lines have half turn symmetry and, in this sense, a degree measure of 180.

To systematically promote integration of turn and figure conceptions of angle with bodily motion, students use flags, ribbon (surveyor tape), and a yardstick to construct and mark rectangular or triangular paths outdoors or in a large indoor space. The use of the yardstick supports further development of unit iteration (as described in Theory of Length Measure level 3). Cones are used to mark each vertex. Constructions are further represented with a set of walking directions. For example, for the first two sides of a rectangular walking path: "Walk straight 5 yards. Turn Right $\frac{1}{4}$ wt (whole turn). Walk straight 8 yards. Turn Right $\frac{1}{4}$ wt." An accompanying diagram shows turn angles and distances, so that re-representation of bodily experience continues to play a prominent supporting role in the development of concepts of angle. For instance, as children walk, how might "straight" be realized in practice? Children usually hit upon walking so that a landmark is kept in sight, resulting in travel at a constant heading. During follow-up discussions, students consider how to modify their directions to create larger (i.e., increase the number of units of length measure for each side) or smaller figures or to generate paths to construct a different shape (e.g., a square).

Generating a walking path for an equilateral triangle challenges students to further split whole-turn units into thirds and to employ these turn angles to walk an equilateral triangle. This is often challenging for students because textbook figures usually represent interior angles, but not exterior or turn angles. Consequently, students often first attempt to turn less than $\frac{1}{4}$ whole turn and are surprised that the resulting path cannot close. From the perspective of the development of rational number, the measure of turn offers opportunities to explore and compare different splits of unit (e.g., $\frac{1}{4}$ whole turn, $\frac{1}{3}$ whole turn).

Further integration of concepts of angle is supported by introducing a new tool, a circular protractor, and a new measure, the degree, as a refinement of the metric of body rotation, the whole turn. These new material and conceptual tools are employed to extend students' investigations of polygon paths. Rather than hand out protractors and direct students in their use, teachers challenge students to reverse-engineer a circular protractor by marking

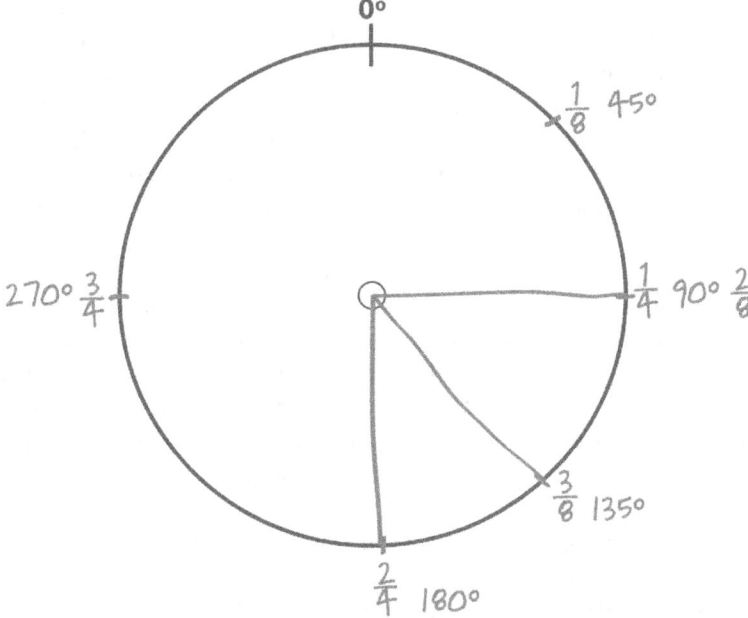

FIGURE 6.7   Reverse-engineering a protractor by partitioning a whole turn and defining a degree

fractions of whole turns on a circle. They explore relations among parts of whole turns (e.g., $\frac{1}{8}$ whole turn = $\frac{1}{2} \times \frac{1}{4}$ whole turn). These part turns are then re-expressed as degrees, with a degree defined as $\frac{1}{360}$ whole turn, so that, for example, 45° is equivalent to $\frac{1}{8}$ whole turn, as is illustrated in Figure 6.7.

Using a pencil with one end at the center of the circle, students enact degree measures by rotating from 0 either clockwise or counterclockwise. Then students use circular protractors to measure angles-as-figures, including supplementary angles, again using a pencil at the center of the protractor to enact the measures as rotations, as illustrated in Figure 6.8.

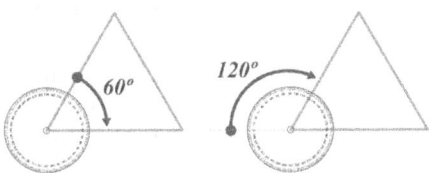

FIGURE 6.8   A turn angle and an interior angle can be measured with a protractor at each vertex of a polygon

After students have been familiarized with the operation of a circular protractor, they use rulers and protractor to re-represent rectangular or triangular paths on paper in degree measure, accomplished by specifying lengths with a ruler and turn angles with the protractor. This challenge of re-representation links bodily motion in large-scale space with imagined and enacted moves and turns on paper. To meet this representational challenge, students must learn procedural skills associated with the practicalities of using protractors to measure magnitudes of angles, such as aligning direction of travel with 0 and centering the protractor on each vertex as it emerges during the imagined walk. Measuring an exterior angle with a protractor during the construction of a triangle often puts to practical use the extension of the sides to make the turn angle more visible by helping students align the start of the turn angle with the 0 measure of the protractor. Students investigate the sum of the measures of the turn (exterior) angle and the interior angle at each vertex of a rectangle and triangle. This sum must be 180°, because together the exterior and interior angles constitute a straight line (an angle-as-figure perspective) or half of a whole turn (an angle-as-turn perspective).

Students next use a straightedge to draw any triangle that they like (see Angle unit 3). They then use rulers and protractors to write path directions for creating the triangle they have drawn, with path moves measured in units of length and turns measured in degrees. We encourage students to consider the sum of the exterior angles of all the triangles drawn and to begin to think about why they must all be exactly the same measure (360°, or 1 whole turn).

Table 6.2 summarizes the conceptual milestones in beginning to integrate the two perspectives of angle-as-turn and angle-as-figure.

Generating and Justifying Angle Measure Theorems

At the fourth level of the angle construct, students generate and justify angle measure theorems, as summarized in Table 6.3. A theorem is a statement about relations among angles that invites explanation of why, so we encourage the development of theorems not as statements to be memorized, but as relations that need to be explained.

For example, although students are often taught that the sum of supplementary angles is 180°, they should also appreciate the necessity of this relationship. That is, they should be able to explain that because straight lines have $\frac{1}{2}$-turn symmetry and $\frac{1}{2}$ turns have a measure of 180°, the measures of angles formed by intersections with a straight line must collectively constitute this sum. An implication of the supplementary angle relation is that the interior angle and the exterior (turn) angle at each vertex of a polygon must sum to 180°,

TABLE 6.2 Integrating Angle-as-Turn with Angle-as-Figure Through Walking Paths

Performances	*Examples*
Demonstrate representational and procedural competence with protractor in two dimensions.	Understands protractor design as partitioning of 1 whole turn and angle measure as amount of turn between two segments. Demonstrates use of 0 as initial heading (e.g., align 0 with one line segment) and then obtains the angle measure in degrees by aligning with the second line segment. (This is first accomplished when one segment is horizontal, then for any orientation of angle.)
Relate openings of figures of angles to amount of rotation that would bring one line segment onto the other.	"If I turn this line segment onto this other one, that's a 45° turn." 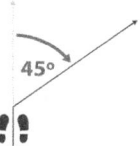
Conceive of 1° as $\frac{1}{360}$th of a whole turn.	"One degree is only one three-hundred-sixtieth of a whole turn. That's a very little amount of turning."
Interpret 3-splits of body turns, such as $\frac{1}{3}$, as measures.	Constructs and identifies three $\frac{1}{3}$ turns for equilateral triangle. (See Figure 6.6 in text.)
Construct rectangles and squares as paths with turns. Compare angles-as-figures using congruency.	Constructs and identifies four $\frac{1}{4}$ turns to either the right or the left for rectangle path. "These angles are congruent. I can slide one vertex to the other vertex, and the angles match."
	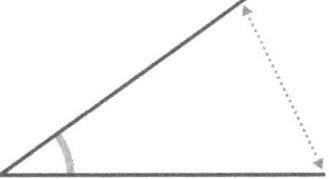 
Regard 2-splits of bodily turns, such as $\frac{1}{2}$, $\frac{1}{4}$, and $\frac{1}{8}$, as measures of magnitude of rotation.	"This is a quarter turn because you are only going one-fourth of the whole way around." "A three-quarter turn is greater than a half turn because you are turning more." Demonstrates difference between $\frac{1}{2}$ turn to the right (clockwise) and $\frac{1}{2}$ turn to the left (counterclockwise).

TABLE 6.3 Generating and Justifying Theorems

Performances	Examples
Generalize total trip theorem to all polygons.	"This regular polygon has 10 sides. So I know that it has to turn the same amount at each vertex. That's $\frac{360}{10}$°, or 36°." "Even though this triangle is not regular, the sum of its turn angles must be 360° because the walker makes one whole turn if she starts and ends pointed in the same direction." "If you add up right and left turns separately and subtract the left turn total from the right turn total, then the total amount of right turn to make this shape (a nonconvex polygon) is still 360°."
Generate and justify total trip patten for familiar polygons, such as squares and triangles. At each vertex, identify and measure interior and exterior angles.	"The walker has to make a 90°, or one-quarter, turn at each corner to stay on the path and come back to its starting point. So the total trip around the shape is 360°, or one whole turn."
Explain, justify, and use vertical (opposite) angle theorem.	"Angle *FEI* and angle *GEH* are congruent because they are both supplements of angle *IEH*." "Angle *FEI* and angle *GEH* are congruent because a half-turn rotation brings *FEI* onto *GEH* and *GEH* onto *FEI*."

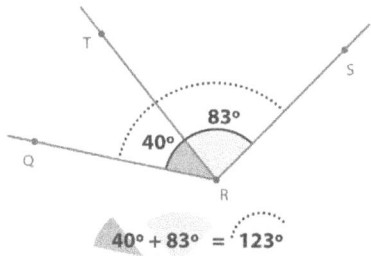

Explain, justify, and use supplementary angle theorem.	"If angle *QRT* is 40° and angle *TRS* is 83°, then angle *QRS* is 123°."

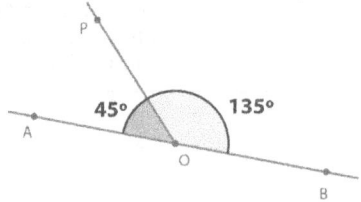

"If angle *AOP* is 45°, then angle *BOP* is 135°, because the sum of the angles must be 180° because *AB* is a straight line, and straight lines have half turn symmetry (180°)."

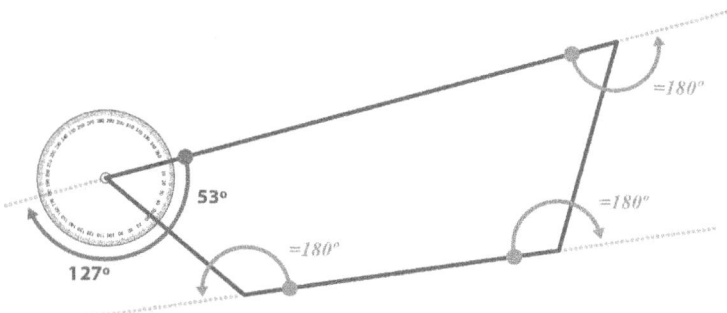

FIGURE 6.9 The sum of the interior and exterior angles at a vertex in a polygon is 180°

as illustrated previously, but this idea is now deployed to generate other polygons, as depicted in Figure 6.9.

Students also explain that the opposite (vertical) angles formed by two intersecting lines are congruent, as shown in Figure 6.10, by appealing to rotational symmetry (the opposite angles can be brought onto one another through a $\frac{1}{2}$ turn about the intersection). Note that symmetries of straight lines are an important adjunct to developing explanations of theorems that are often otherwise simply memorized and then forgotten.

As students become more familiar with these theorems, they recognize, first for cases of familiar polygons and then more generally, that the sum of the turn (exterior) angles of any convex polygon must be 360° (the 360° theorem, Abelson & diSessa, 1983), as shown in Figure 6.11. Rather than accepting this theorem as grounded in empirical experience alone, they further justify this generalization by appealing to the one entire rotation completed by beginning at and returning to the same heading as an agent "walks" along the edges (sides) of a polygon, so that, as students often put it, polygons are "circles."

Students use circular protractors and rulers to construct other polygons as paths with line segments and turns (exterior angles). They usually begin

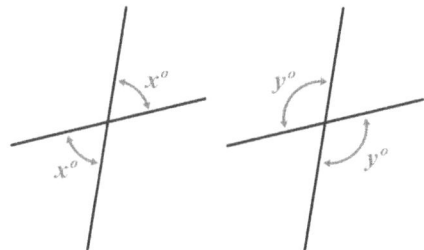

FIGURE 6.10 Opposite angles of intersecting lines are congruent, as shown by rotation symmetry

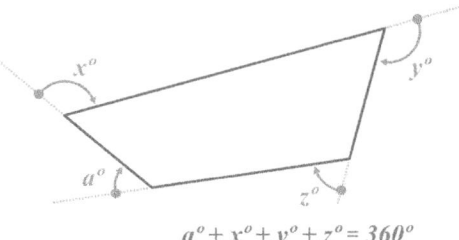

$$a° + x° + y° + z° = 360°$$

FIGURE 6.11 The sum of the turn angles for a convex polygon is 360°

with the path for a regular pentagon and subsequently develop the path for an irregular pentagon. As Figure 6.12 illustrates, the 360° theorem also holds for nonconvex polygons (that is, those with at least one interior angle greater than 180°), if conventions can be adopted about positive and negative turn angles. Class conversation focuses on conjectures that explain each angle sum, and the sums of turn angles and interior angles are compared across different polygon-path constructions.

Developing New Understandings of Figures and Structures via Angle Theorems

As students continue to explore angles, they use the total trip (360°) theorem and their knowledge of other properties of figures, such as symmetries, to establish new theorems and properties. For example, at each vertex of a triangle, the sum of the turn angle and the interior angle is 180°, so $TA_1 + IA_1 = 180°$ and $TA_2 + IA_2 = 180°$ and $TA_3 + IA_3 = 180°$. So, $TA_1 + IA_1 + TA_2 + IA_2 + TA_3 + IA_3 = 540°$. Because the sum (540°) is the same regardless of the order of addends, we can say $(TA_1 + TA_2 + TA_3) + (IA_1 + IA_2 + IA_3) = 540°$. But we know that the sum of the turn angles must be 360°. This means that the sum of the interior angles of a triangle must be 180°. Similar approaches

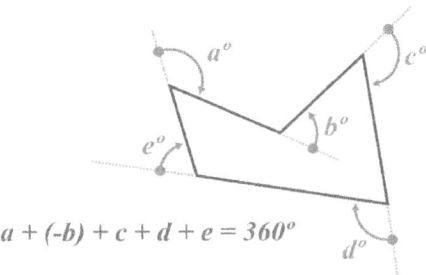

$$a + (-b) + c + d + e = 360°$$

FIGURE 6.12 Clockwise turn angles are assigned a positive value, counterclockwise ones (angle *b*) a negative value. The sum of angles is 360°

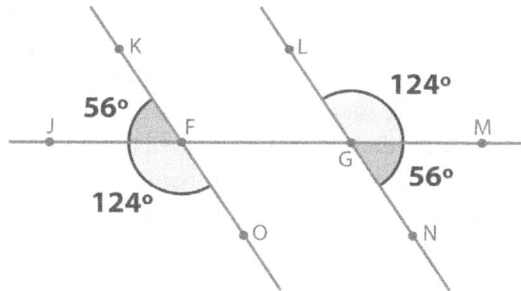

FIGURE 6.13   Properties of angles made by transversals of parallel lines

can be developed for quadrilaterals, and it is often interesting to students that even quadrilaterals that are not convex are nonetheless still governed by the 360° theorem.

Symmetries can also be used to justify properties of angles made by transversals of parallel lines. Transversals are lines that intersect two or more parallel lines. For example, Figure 6.13 depicts a transversal of parallel lines *LN* and *KO* by *JM*. If one chooses the midpoint of the distance between the two parallel lines (the midpoint of *FG*) as the center of rotation, then half-turn rotation brings angle *JFK* onto angle *MGN*, so we know that these angles are congruent and therefore must have the same degree measure.

Finally, students extend these ideas about path and angle into the wider context of wayfinding. They use an orienteering compass to set a heading and then use paces and the compass to write directions that create familiar paths, such as squares and rectangles, but now with degree measure defined by an external frame of reference (magnetic North). Paces are converted to feet to facilitate comparisons among directions. Using distance and directions and beginning from unique starting points, students make a map that shows the locations of a set of objects designated by flags. Students' maps are compared, and the teacher highlights the virtue of a common origin, the need to re-represent feet or yards with a scale (such as 1 in = 5 ft), and the convention of orienting maps so that "up" is North. Students are introduced to polar coordinates as a solution for generating a map with a common origin and agreed-upon conventions for representing distance and direction. Students use their maps to solve wayfinding problems in large-scale space (Lehrer & Pritchard, 2002).

The five levels of benchmark understanding that we have described for angle span from noticing everyday instances of "bends" or "pointy" to inventing and using tools (e.g., protractors, maps, orienteering compasses) to explore and generalize mathematical theorems and relations. The work with mathematical theorems supports students' efforts to knit the logical connections between the many apparently different instantiations of angle. Wayfinding investigations help students extend concepts of angle as rotation to an external frame of reference (magnetic North). This common frame of reference is abstracted to describe location in large-scale space by a polar coordinate system.

Chapter 8 describes research on students' learning of these concepts about angle. The examples were drawn from task-based interviews conducted with a sample of children, summative assessment items given to all student participants, and formative assessment conversations conducted by teachers in their classrooms. Together, these findings provide a perspective on children's understanding of angle as it evolves during supporting instruction across grades one through five.

References

Abelson, H., & diSessa, A. (1983). *Turtle geometry. The computer as a medium for exploring mathematics.* Cambridge, MA: MIT Press.

Bryant, P. (2009). Understanding space and its representation in mathematics. In T. Nunes, P. Bryant, & A. Watson (Eds.), *Key understandings in mathematics learning.* London, UK: Nuffield Foundation.

Clements, D. H., & Battista, M. T. (1990). The effects of Logo on children's conceptualizations of angle and polygons. *Journal for Research in Mathematics Education, 21,* 356–371.

Clements, D. H., Battista, M. T., Sarama, J., & Swaminathan, S. (1997). Development of students' spatial thinking in a unit on geometric motions and area. *The Elementary School Journal, 98*(2), 171–186.

Clements, D. H., & Burns, B. A. (2000). Students' development of strategies for turn and angle measure. *Educational Studies in Mathematics, 41*(1), 31–45.

Crompton, H. (2015). Understanding angle and angle measure: A design-based research study using context-aware ubiquitous learning. *International Journal for Technology in Mathematics Education, 22*(1), 19–30.

Devichi, C., & Munier, V. (2013). About the concept of angle in elementary school: Misconceptions and teaching sequences. *The Journal of Mathematical Behavior, 32*(1), 1–19.

Henderson, C. W., & Taimina, D. (2001). *Experiencing geometry in Euclidean, spherical, and hyperbolic spaces* (2nd ed.). Upper Saddle River, NJ: Prentice Hall.

Kaur, H. (2013). Children's dynamic thinking in angle comparison tasks. *Proceedings of PME 37, 3,* 145–152.

Kaur, H. (2020). Introducing the concept of angle to young children in a dynamic geometry environment. *International Journal of Mathematical Education in Science and Technology, 51*(2), 161–182.

Keiser, J. M. (2004). Struggles with developing the concept of angle: Comparing sixth-grade students' discourse to the history of the angle concept. *Mathematical Thinking and Learning, 6*(3), 285–306.

Latsi, M., & Kynigos, C. (2011). Meanings about dynamic aspects of angle while changing perspectives in a simulated 3-D space. *Proceedings of PME 35, 3,* 121–128.

Lehrer, R., Guckenberg, T., & Lee, O. (1988). A comparative study of the cognitive consequences of inquiry-based Logo instruction. *Journal of Educational Psychology, 80*(4), 543–553.

Lehrer, R., Jenkins, M., & Osana, H. (1998). Longitudinal study of children's reasoning about space and geometry. In R. Lehrer, & D. Chazan (Eds.), *Designing learning environments for developing understanding of geometry and space* (pp. 137–167). Mahwah, NJ: Lawrence Erlbaum Associates.

Lehrer, R., & Pritchard, C. (2002). Symbolizing space into being. In K. Gravemeijer, R. Lehrer, B. van Oers, & L. Verschaffel (Eds.), *Symbolization, modeling and tool use in mathematics education* (pp. 59–86). Dordrecht, Netherlands: Kluwer Academic Press.

Lehrer, R., Randle, L., & Sancilio, L. (1989). Learning pre-proof geometry with Logo. *Cognition and Instruction, 6*, 159–184.

Masuda, Y. (2009). Exploring students' understanding of angle measure. In M. Tzekaki, M. Kaldrimidou, & H. Sakonidis (Eds.), *Proceedings of the eighth annual meeting of the North American chapter of the International Group for the Psychology of Mathematics Education* (pp. 169–177). East Lansing: Michigan State University.

Mitchelmore, M. C. (1998). Young students' concepts of turning and angle. *Cognition and Instruction, 13*(3), 165–284.

Mitchelmore, M. C., & White, P. (2000). Development of angle concepts by progressive abstraction and generalization. *Educational Studies in Mathematics, 41*(3), 109–238.

Olson, D. R. (1996). *Cognitive development: The child's acquisition of diagonality.* New York, NY: Psychology Press.

Papert, S. (1980). *Mindstorms: Children, computers, and powerful ideas.* New York, NY: Basic Books.

Wilson, P. S., & Adams, V. M. (1992). A dynamic way to teach angle and angle measure. *Arithmetic Teacher, 39*(5), 6–13.

7

MEASUREMENT MODELS OF ARITHMETIC OPERATIONS AND RATIONAL NUMBER

Fractions are pervasive in K–5 mathematics education, but children and teachers in the elementary grades often find fractions challenging to construct and interpret and, further, find it difficult to make sense of arithmetic operations on fractions. These difficulties stem, at least in part, from the fact that what students and teachers have learned about whole numbers and arithmetic operations on whole numbers can be misleading when applied to fractions (Depaepe et al., 2015; Ni & Zhou, 2005; Siegler & Lortie-Forgues, 2017). Moreover, instruction about fractions in elementary mathematics too often primarily emphasizes how to carry out procedural rules for solving problems, rather than providing a conceptual frame for situating fraction knowledge within a wider system of mathematical knowledge (e.g., Smith, 2002).

Historically, fractions arose in response to the need to refine measures of magnitudes by fracturing units (Davydov & Tsvetkovich, 1991), subdividing them into ever-finer gradations to meet particular requirements of different situations of measure. For students, the need to refine measurements first arises when they confront challenges of measuring magnitudes of length, area, angle, and volume that are not divisible by whole-number units. As described in Chapters 3–6, subdividing units challenges students to conceptualize relations between magnitudes of a subdivided unit and its parent unit, to measure magnitudes in subdivided units, and to engage in additive and multiplicative comparisons of these measures. In the current chapter, we lift away from this embedding of rational number within the strands of measure to illuminate how children's investigations of measurement can support their understandings of fractions and arithmetic operations involving fractions. Accordingly, this chapter revisits students' ways of thinking about measure, but now highlights measure as an incubator of conceptions about rational number.

DOI: 10.4324/9781003287476-7

We describe three major ways in which the development of rational number can be both provoked and supported as students engage in processes of measure. First, we describe how conceptions of measure that arise during the course of constructing whole-number unit measurements can serve as foundations, rather than as impediments, to subsequent learning about fractions. Second, we trace a developmental trajectory for interpreting fractions as measured quantities and illustrate how arithmetic operations of addition and subtraction can be employed to compare these quantities. As noted, this strand originates in the need to fracture units to refine a measure of the magnitude of a length, area, volume, or angle. The third, parallel line of development unfolds as students come to regard fractions as operators on measured quantities, with comparisons of quantities grounded in operations of multiplication and division. This tripartite organization is grounded in a wide-ranging body of related research that articulates children's development of quantitative reasoning (Smith & Thompson, 2007; Thompson, 1993, 2011; Thompson & Carlson, 2017), fraction schemes (Steffe & Olive, 2010; Thompson & Saldanha, 2003), and forms of partitioning and reasoning about covariation described as splitting (Confrey, Maloney, Nguyen, & Rupp, 2014; Confrey & Smith, 1995). We also draw on research that explores how students' understandings of length measure can support learning about fractions as measured quantities and as operators on measured quantities (Cortina, Višňovská, & Zúñiga, 2014; Lehrer & Pfaff, 2011; Lehrer & Slovin, 2014). In the sections that follow, we focus primarily on length measure to illustrate each of the major lines of the development of rational number that are supported by instruction in measure, but measure of area or volume or angle also provides ample opportunities to ground learning about rational number.

Initial Resources for Reasoning About Rational Number in Measure

Ways of thinking about whole-number measure can serve as building blocks to support the development of reasoning about fractional measure and about fractions as numbers that operate on these fractional measures. These initial resources do not arise spontaneously but are developed with instructional support, as described in the curricular units associated with each of the domains of measure. In this section, we highlight some of these initial contributors to reasoning about rational number.

Unit Iteration

A magnitude of length measured in a whole number of units is constructed by translating and accumulating a unit length. The magnitude of the measured length is compared to the magnitude of the unit length, as in a 5-ft length is five times as long as 1 ft. The implicit ratio of a measure of 5 ft can be made more visible to children by systematically using the language *times as long* and by consistently coordinating that phrase both with its symbolic expression (as in 5 ft = 5 × 1 ft) and with literal enactment (e.g., five iterations or five copies of a unit length). Iteration can be extended to composite units, as in 10 ft is five times as long as 2 ft. Reasoning about iterations of composite units supports coordination of multiple levels of unit—for instance, understanding that 10 ft results from only five iterations of 2 ft because 2 ft is simultaneously a unit-of-units and composed of separate 1-ft units. Unit iteration and units coordination combine to support subsequent reasoning about fractional quantities as constructed by the translation of fractional units. For example, three copies of $\frac{1}{2}$ u construct a distance with measure $\frac{3}{2}$ u, where $\frac{1}{2}$ u is understood simultaneously as the unit of measure and as defined by its relation to 1 u.

Length Measure as a Point Along a Path

As explained in earlier chapters, the iterative perspective on length measure is complemented by a metaphor of motion in which length is conceived of as a continuous path that is traveled from a starting point to an ending point. A continuous perspective on whole-number measure sets the stage for marking distances traveled along a path with units that are not whole numbers. From this perspective it makes sense to indicate a measure of a length that has not yet "arrived" at a whole number of units along a path, both for distances between 0 u and 1 u and for distances that exceed 1 u, such as $\frac{3}{2}$ u.

Measure-Magnitude Distinction

A magnitude refers simply to spatial extent—an amount of length, area, volume, or angle. The measure of a fixed magnitude of length varies with unit, a recognition that emerges as young children re-measure the same magnitude of length with different units. Familiarity with this distinction helps students recognize that the same distance can be measured with a unit length and with a part-unit length and note that the measured quantity will be greater with the part-unit length than with the unit length. This measure-magnitude distinction also helps students anticipate relations among part units, so that if, for example, the part unit is one-fourth times as long as the unit, the number of copies needed will increase over what it is with the half-unit or unit metrics. This recognition helps students order the magnitudes of fractional-unit lengths and to make sense of relations like $\frac{3}{2}$ u = $\frac{6}{4}$ u, where the equality refers to distance traveled.

Symbolizing Measure

Although whole-number measure begins in activity, symbolization fixes and transports unit iteration and distance traveled to new contexts of measure. Important steps in symbolization include representing the origin as 0 and coordinating counts of units with the endpoints of each unit by representing these endpoints with numerals, such as 1, 2. As students learn about measure in part units, symbolization serves as an important mediator to help them keep track of levels of unit and of multiple labels for the same distance along a path, as in $\frac{3}{2}$ ft and $1\frac{1}{2}$ ft. In this instance, the double label indicates coordination of measure in split unit and unit (where 1 ft is recognized as the composite $\frac{2}{2}$ ft), which constitutes evidence of the initial steps in the construction of what Steffe and Olive (2010) call an iterative fraction scheme.

Additive and Multiplicative Comparison

Students employ arithmetic operations of addition or subtraction to compare whole-number quantities. For length measure, joining or gluing two or more lengths together constructs a length with a measure of the sum of both measured lengths, a u + b u. Finding a difference in measured lengths by superimposing them at a common starting point is represented as subtraction; the noncongruent portion corresponds to the result of a u − b u. These operations allow students to compare lengths additively, to answer questions concerning *how much more or less than*. Multiplicative comparisons between unit and measured quantity are signaled by *times as long,* and multiplicative relations are tested with unit and composite-unit iteration. Measured quantities are also compared in this way; for example, a length with measure 12 u is three times as long as a length with measure 4 u. Prior experiences with additive and multiplicative comparisons are important resources for extending these systems of comparison to fractional quantities.

Rational Numbers as Measured Quantities

In this section we describe conceptual facets that become increasingly coordinated as students refine measures, here with a focus on length, by creating unit splits with factors of 2 (e.g., $\frac{1}{8}$ u) and 3 (e.g., $\frac{1}{3}$ u), compositions of 2 and 3 (e.g., $\frac{1}{6}$ u), and eventually any arbitrary split, as in 5 or 10. Hand in hand with these developments come corresponding developments in thinking about fractions as actors (operators) on measured quantities. Nevertheless, for clarity, we will discuss these developments separately and in the next section focus only on the development of measured quantity.

Two-Split of a Unit Length and Half-Unit Iteration

Although fracturing units arises from the need to refine measures of magnitude, not all fractures are equally apt entrées for students. An early-developing analog system of magnitude representation (Beck, 2015) enables kindergarten and first-grade children to engage in multiplicative comparisons of halving and doubling, both for collections of objects and for lengths, without counting or measuring (Barth, Baron, Spelke, & Carey, 2009). This early intuitive knowledge, as well as the ubiquity of organisms and objects in children's worlds that have bilateral symmetry, suggests that a 2-split of a unit is a productive initial step toward the refinement of measure. By a *2-split* we mean a split that results in the production of two congruent parts. Moreover, 2-splits can be applied recursively to generate a number of parts that increases exponentially (Confrey & Smith, 1995). Such equipartitioning is a foundation of rational number (Confrey, et al., 2014).

Conceiving of fractional quantities as measured in a 2-split unit is initiated by staging situations that provoke the need to measure a length that falls at the midpoint of a familiar unit, such as a person-foot. Young students readily call the distance "half" of the unit, and the purpose of the instruction is to enlarge and refine this initial conception of halving. To do so, we engage students in the refinement of a whole-number tape measure they have previously constructed. The new challenge posed is to indicate the distance traveled from 0, but now in half-units. To find a half-unit, students fold a paper-strip unit length into two congruent parts, so that the ends of the paper "match." Students tear off one of the two parts, designate the part as a half-unit, and iterate it two times. Other noncongruent 2-parts of the unit are also iterated two times to establish that only the part constructed by the equipartition, the "matching part," can be copied two times to construct a length just as long as one unit. (Two copies of any noncongruent part are either longer than or shorter than one unit length.) This realization establishes a multiplicative relation, that the unit length is two times as long as the half-unit. The relation can be tested by iterating (copying) the half-unit, so that iteration can be viewed as a multiplicative relation, just as it is for whole numbers of units.

Teachers then introduce the symbolization of the half-unit as $\frac{1}{2}$ u. The "bottom number" signifies the number of congruent parts that resulted from the fracture of the unit. The "top number" signifies the number of copies of one of these congruent fractures. Thus, $\frac{1}{2}$ u refers to one copy of a 2-split part unit. Students revisit the unit length that results from two copies of the split unit with an emphasis now on its measure as $\frac{2}{2}$ u. Measuring the same magnitude of length both as 1 u and as $\frac{2}{2}$ u restates the measure-magnitude distinction first experienced with different whole-number units of measure and provides an entrée into an iterative fraction scheme (Steffe & Olive, 2010). The metaphor of travel is particularly critical to investigating iterations that result in measured lengths longer than one unit, as in $\frac{7}{2}$ u. Coordinating measure in *unit* and *half-unit,* as in 3 u and $\frac{6}{2}$ u, is supported by double labeling of interval endpoints on a tape measure. Teachers often unpack a double-labeled tape measure to reveal the separate counting

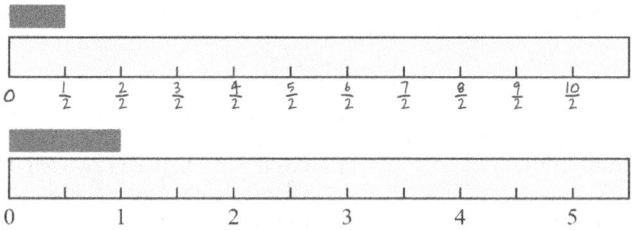

FIGURE 7.1 Coordinating measures of length in unit and half-unit (grade 1)

schemes, as illustrated in Figure 7.1. In this first-grade class, the teacher initially juxtaposed two congruent blank tapes. Then she and the students iterated and marked intervals of $\frac{1}{2}$ pf (pink foot) on the top tape measure, followed by iterating and marking intervals of 1 pf on the bottom tape measure. After a common starting point of 0 had been established, the challenge posed to students was to label each interval and to establish correspondences in these different units for the same distance from the origin. Using the tape measure with intervals of $\frac{1}{2}$ pf to locate corresponding midpoints on the unit-interval tape measure, first graders were challenged to relate half-unit measures such as $\frac{5}{2}$ pf to the corresponding unit measure, such as $2\frac{1}{2}$ u. It was helpful for the first graders to interpret measures like $2\frac{1}{2}$ u as 2 u $+ \frac{1}{2}$ u. This notation capitalized on addition as gluing measured lengths or as continuation of motion, as in "We traveled two pink feet and then another half."

Different distances traveled by moving along the paths in Figure 7.1 were compared additively by the first-grade children, as in $\frac{9}{2}$ u $- \frac{5}{2}$ u $= \frac{4}{2}$ u, and $\frac{4}{2}$ u $+ \frac{5}{2}$ u $= \frac{9}{2}$ u. Teachers encouraged students to consider the meaning of the notations as these comparisons were symbolized. For instance, they asked, "Why is it that the bottom numbers (denominators) were the same, but the top numbers (numerators) were not?" Teachers also had students join lengths of different measure and then write symbolic expressions that captured this action and the resulting total distance traveled. Table 7.1 displays indicators of conceptions of fractional quantity that originate in 2-split unit measure of a length and comparisons of 2-split measured lengths.

Four- and Eight-Splits of a Unit Length and Measures in Fourth-Unit and Eighth-Unit

Following investigations with half-unit measures, students go on to consider more refined measures of length made possible by recursive partitioning by two. Repeated partitioning by two is articulated more completely in the third, operator strand, but its importance here is that new part-unit measures are now possible. The first repetition of the splitting process is expressed informally as "half of a half-unit," accompanied by successive 2-splits of a unit

TABLE 7.1 Conceiving of Fractional Quantities as Measured in a 2-Split Unit

Description	Examples
Compares half-unit quantities as differences (subtraction) or as joined lengths (addition).	Expresses length measure as sum ($\frac{1}{2}$ u + $\frac{1}{2}$ u + $\frac{1}{2}$ u = $\frac{3}{2}$ u). Finds difference between two length measures, as in $\frac{7}{2}$ u − $\frac{3}{2}$ u = $\frac{4}{2}$ u (= 2 u).
Establishes equivalence between mixed number and fractional distances traveled (e.g., $\frac{5}{2}$ u = $2\frac{1}{2}$ u).	"I traveled $\frac{5}{2}$ u and that's the same distance as $2\frac{1}{2}$ u." Note: Student may count by half-units and by units but not coordinate the two counts, so check to see if student can relate unit measure to half-unit measure, as in the example.
Iterates by $\frac{1}{2}$ unit and accumulates to generate quantities greater than 1 unit; symbolizes accumulated quantity as, for example, $\frac{5}{2}$ u and says, "five halves."	Using unit-length paper strips creased at every $\frac{1}{2}$ unit and taped together, child travels and, at each crease, symbolizes/writes $\frac{1}{2}$ u, $\frac{2}{2}$ u, $\frac{3}{2}$ u and associates with number word sequence one-half u, two halves u, three halves u.
Symbolizes one half-unit as $\frac{1}{2}$ u and interprets it as a quantity—a distance traveled from the origin. Interprets denominator as the number of congruent parts (two) that the unit was split into. Interprets numerator as the number of copies of one of these two congruent parts.	Interprets $\frac{2}{2}$ u as two copies of $\frac{1}{2}$ u and demonstrates with two iterations of $\frac{1}{2}$ u. Folds one unit length into two congruent parts and tears off one part. Points to remaining part and labels it as $\frac{1}{2}$ unit. Travels with fingers from 0 to midpoint of 1 unit, identifying distance traveled as $\frac{1}{2}$ unit. "The bottom number means that we chopped the HF (a teacher's-foot strip) unit into two equal parts." "The top number means that we have one copy" (points to $\frac{1}{2}$-unit length).
Establishes that a 2-split partitions a unit length into two congruent parts (equipartitions). Two iterations of one half-unit length construct the unit length.	"That's not one-half because the two parts are not the same length." "That is one-half because the parts match" (shows congruence). Demonstrates that two iterations of "one-half" are the same distance traveled as 1 unit length. Demonstrates that two iterations of a 2-partition other than one-half are either less than or more than 1 unit.

length to produce four congruent partitions. The conceptual terrain initiated with a single 2-split is revisited, but now with $\frac{1}{4}$ u. For example, four copies of $\frac{1}{4}$ u are exactly the same length as 1 u, nine copies of $\frac{1}{4}$ u construct the same distance traveled as $2\frac{1}{4}$ u, and the sum of $\frac{2}{4}$ u and $\frac{5}{4}$ u constructs a length with measure $\frac{7}{4}$ u. Four-split units are related to 2-split units, in part by traveling distances along constructed tape measures and finding correspondences among measures in unit, half-unit, and quarter-unit, and in part by considering problems like these: "Which is longer: $\frac{3}{2}$ u + $\frac{5}{2}$ u or $\frac{3}{4}$ u + $\frac{5}{4}$ u? How much longer? How many times as long?"

Recursive partitioning is continued to produce an 8-split as "half of half of a half-unit," and the terrain of part-unit quantities is expanded to include eighths and, subsequently, continued recursions to produce 16, 32, ... parts. Through paper folding and symbolizing the acts of folding, students achieve tangible access to a simple exponential structure, 2^n, with a common ratio (here, 2). A signature of student understanding of this structure is that they can anticipate the result—that is, the number of parts that will be generated by successive 2-splits—so that thinking of ever-finer subdivisions of a unit is tangible and accessible. Thinking in this way is an informal avenue to considering fractions as infinitely dense, because one can always imagine a successor that is half as long as its predecessor. The 2-split structure is also a handy resource for other investigations, as when students think of movement of an area measure through a length measure that can be subdivided as finely as one likes to generate a measured volume (as described in Chapter 5 of this volume).

Three-Splits and Compositions of Two- and Three-Splits of a Unit Length

A common unit of length measure in the United States equipartitions a unit length, the foot, into 12 congruent parts, each called an inch. Consequently, we introduce 3-splits and compositions of 2- and 3-splits by engaging students in reverse-engineering the structure of a standard foot ruler. In this standard tool, the inch is recursively 2-split, most commonly into eighths and even sixteenths, but only composition of 2- and 3-splits will produce a twelfth of a foot. (An arbitrary twelfth is easily produced by iterating any length 12 times and then designating the constructed length as 1 unit, but here a standard unit length must be equipartitioned.) We provide students with a length that we call a Big Foot (BF), which is twice as long as the standard 12-in foot, and challenge them to partition the Big Foot into 12 Large Inches (LI). BF is similar to the foot, but to construct it, students must generate and coordinate partitions of the unit rather than simply copying a standard ruler. Supporting investigations include 3-splitting a BF to produce $\frac{1}{3}$ BF, recursive 3-splitting with an exponential structure governed by 3 (3, 9, 27, ...), compositions of 2- and 3-splits to generate $\frac{1}{6}$ BF and $\frac{1}{3}$ BF, and symbolizing the same distance from the origin with different unit measures, such as $\frac{1}{2}$ BF, $\frac{3}{6}$ BF, and $\frac{6}{12}$ BF (see, for example, curriculum Length unit 8). Table 7.2 summarizes some of the conceptual underpinnings of constructing fractional quantities with these new fractures of a unit length.

With these foundations, it becomes feasible to explore other unit equipartitions, such as fifths and tenths, that are less amenable to exploration via paper folding. These equipartitions are also symbolized and interpreted to extend students' sense of everyday fractional quantities of measure. We have usually introduced fifths and tenths in contexts of the metric system of measure.

TABLE 7.2 Comparing Quantities Constructed by 3-Splitting and by Compositions of 2 and 3

Description	Examples
Compares $\frac{1}{3}$ u or $\frac{1}{6}$ u or $\frac{1}{12}$ u quantities as differences (subtraction) or as joined lengths (addition).	Expresses length measure as sum ($\frac{1}{3}$ u + $\frac{3}{6}$ u = $\frac{2}{6}$ u + $\frac{3}{6}$ u = $\frac{5}{6}$ u). Expresses and finds differences between measured lengths ($\frac{10}{3}$ u − $\frac{4}{3}$ u = $\frac{6}{3}$ u = 2 u). Uses knowledge of equivalent measures to find sums and differences with unlike units ($\frac{1}{2}$ u + $\frac{1}{6}$ u + $\frac{1}{3}$ u = $\frac{3}{6}$ u + $\frac{1}{6}$ u + $\frac{2}{6}$ u = $\frac{6}{6}$ u).
Establishes equivalence among quantities measured in third-units, sixth-units, and twelfth-units.	"I traveled $\frac{3}{2}$ u, and that is the same distance traveled as $\frac{9}{6}$ u and $\frac{18}{12}$ u."
Establishes equivalence between mixed number and fractional distances traveled (e.g., $\frac{10}{3}$ u = $3\frac{1}{3}$ u).	"I traveled $\frac{5}{3}$ u, and that's the same distance as $1\frac{2}{3}$ u."
Iterates by $\frac{1}{3}$ unit or $\frac{1}{6}$ unit or $\frac{1}{12}$ unit and accumulates to generate quantities greater than 1 unit; symbolizes accumulated quantity as, for example, $\frac{10}{3}$ u.	Using unit-length paper strips creased at every $\frac{1}{3}$ unit, student travels and counts at each crease, $\frac{1}{3}$ u, $\frac{2}{3}$ u, $\frac{3}{3}$ u, $\frac{4}{3}$ u, … .

Fractions as Operators on Measured Quantities

The fraction-as-operator strand attends to how fractions transform measured quantities. This strand is intended to refine and elaborate students' informal conceptions of splitting units by reconceptualizing these as scalar multiplication of measured lengths. A scalar multiple is a number that acts on a measured length to stretch or shrink it or, in the special case of a number equivalent to 1, leaves the length unchanged. For example, in $\frac{1}{2}$ × 5 cm, the $\frac{1}{2}$ acts on the 5 cm by halving it, so the length now has a measure of $\frac{5}{2}$ cm or $2\frac{1}{2}$ cm. The scalar multiple creates a length one-half times as long as the starting length, and the starting length is two times as long as the product length. Hence, *fraction-as-operator* grounds multiplicative comparisons between lengths and eventually culminates in the generalization of a reciprocal relation between A measured in B and B measured in A (this generalization is reflected in the highest level in the length construct). The fraction-as-operator strand also includes dynamic generation of fractional areas and volumes, as described in Chapters 4 and 5, and these explorations afford opportunities to consider multiplicative comparisons (e.g., as in doubling or halving an area or volume). In the sections that follow, we highlight forms of instruction that help students develop conceptions of fractions as actors. These conceptions complement and enable conceptions of fractions as measured quantities.

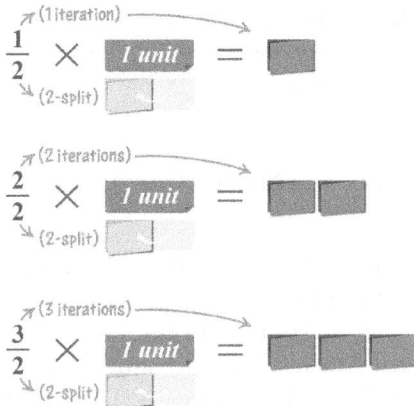

FIGURE 7.2 Split-and-copy interpretation of the product of a scalar multiple and a unit length

Initial Steps in Developing a Sense of Fraction as Operator

Young students' familiarity with halving and their experience in the material world of generating and testing equipartitions of a unit length to satisfy the criterion of $\frac{1}{2}$ u are resources for recontextualizing the act of folding paper unit lengths as two coordinated actions, splitting and copying, that together generate a product. The act of creasing the paper creates a 2-split, embodied by a hand gesture of chopping and then tearing off (or folding) a part, to highlight its status as another unit of measure, and followed by copying (iterating) the part unit to constitute a measure in $\frac{1}{2}$ u, as illustrated in Figure 7.2.

The split-copy interpretation of the action of the scalar multiplier extends to quantities greater than 1 u, as Figure 7.3 illustrates.

Teachers support student investigations that establish the commutativity of the dual actions. We have been describing splitting to create a part unit and then copying, but copying can precede splitting to construct the product length. For example, in Figure 4.3, five copies of 2 u, followed by a 2-split of 10 u. To complement the additive comparisons emphasized in the second strand, students are encouraged to extend whole-number multiplicative comparisons to fractional quantities that have been constructed by 2-splitting and copying. For instance, the $\frac{5}{2}$ (2 u) depicted in Figure 7.3 is five times as long

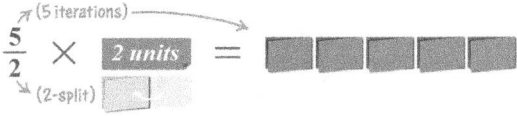

FIGURE 7.3 Split-and-copy interpretation of the product of a scalar multiple and a measured length

as $\frac{1}{2}$ (2 u), a result that can be tested by iterating $\frac{1}{2}$ (2 u), or 1 u, five times. (Initially students do not simultaneously recognize that this comparison means that 1 u must be $\frac{1}{5}$ times as long as 5 u.)

Extending the Reach of Fraction-as-Operator to Refine Measure

Working again to refine and elaborate students' intuitions about successive halving, we extend the split-copy metaphor to compositions of 2-splits, emphasizing recursive dual action, as in first constructing $\frac{1}{2}$ u by actions of splitting and copying and then acting on $\frac{1}{2}$ u in the same manner to generate $\frac{1}{4}$ u, a 4-split of 1 u, as depicted in Figure 7.4. The action sequence and product are symbolized as $\frac{1}{2} \times (\frac{1}{2} \times 1$ u$) = \frac{1}{4}$ u. A heuristic to help keep track of the measured quantity is to represent it in the rightmost position and then to act from right to left, as in function notation.

Variations on split-copy include cases like those displayed in Figure 7.5, in which the measured quantity exceeds 1 u and the operator fraction is expressed in fourths. Children are encouraged to interpret $\frac{9}{4} \times 2$ u $= \frac{9}{2}$ u as 4-splitting 2 u into four congruent parts, each of length $\frac{1}{2}$ u, and then constructing the product as nine copies of $\frac{1}{2}$ u, or $\frac{9}{2}$ u. An alternative way to interpret $\frac{9}{4} \times 2$ u is to first iterate (copy) 2 u nine times and then 4-split the resulting 18 u length to form the $4\frac{1}{2}$ u product length. Expressing the same magnitude of length in unit length or half-unit length helps students develop a commutative perspective—for example, seeing $9 \times (\frac{1}{4} \times 2$ u$)$ as equivalent to $\frac{1}{4} \times (9 \times 2$ u$)$.

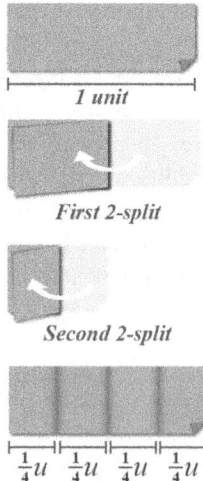

1 unit

First 2-split

Second 2-split

$\frac{1}{4}u$ $\frac{1}{4}u$ $\frac{1}{4}u$ $\frac{1}{4}u$

FIGURE 7.4 Composing 2-splits of a unit length to refine measure

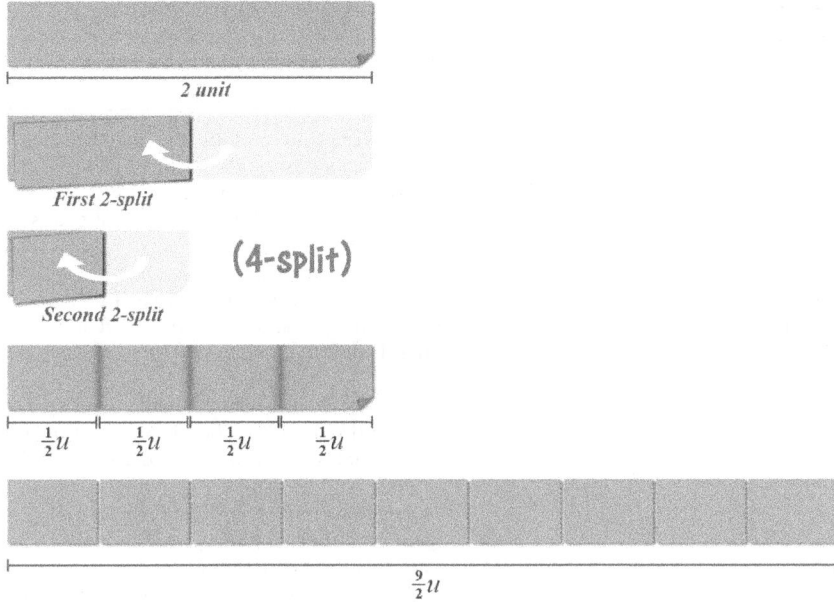

FIGURE 7.5 Four-splitting of measured quantities with length measurement greater than one unit

This recursive construction by split-copy of a measured quantity is extended to eighths so that students have ways of understanding how scalar multiples with factors of two can transform a length. As noted earlier, extending the 2-splitting process helps students understand why fractional quantities are dense. These conceptual underpinnings are described in Table 7.3.

As noted previously for quantities involving 3-splits and compositions of 2- and 3-splits, the conceptual framework outlined in Table 7.3, with its reliance on a split-copy interpretation of scalar multiplication, is extended with student investigations of 3-splits (e.g., $\frac{4}{3} \times b$ u), compositions of 3-splits (e.g., $\frac{1}{3} \times \frac{1}{3} \times b$ u), and compositions of 2- and 3-splits. The copy-split metaphor and conceptual framework are further extended by considering scalar multiples of lengths measured in fractional quantities. For example, an upper-grade student reasoned about $\frac{4}{3} \times \frac{9}{12}$ ft $= \frac{12}{12}$ ft, or 1 ft, "because $\frac{1}{3}$ of $\frac{9}{12}$ is $\frac{3}{12}$ ft, and four copies of $\frac{3}{12}$ ft is $\frac{12}{12}$ ft."

Extending Multiplication and Multiplicative Comparisons

There are two additional contexts of measure that extend students' senses of the meaning of multiplication involving fractional quantities and of multiplicative comparisons. The first involves reasoning about fractional quantities of area and volume. As described in Chapter 4, movement of a length through

TABLE 7.3 Facets of Conceptions of Fractions as Operating by 2-Splitting and Copying

Description	Examples
Compares fractional quantities of composed 2-splits multiplicatively (but not necessarily reciprocally).	"$\frac{3}{2}$ u is three times as long as $\frac{1}{2}$ u." (But not $\frac{1}{2}$ u is $\frac{1}{3}$ times as long as $\frac{3}{2}$ u.) "$\frac{8}{8}$ u is four times as long as $\frac{2}{8}$ u" (demonstrates that four iterations of $\frac{2}{8}$ u is as long as $\frac{8}{8}$ u). "$\frac{8}{4}$ u is eight times as long as $\frac{1}{4}$ u."
Extends 2-split composition to compose $\frac{1}{8}$ u. Interprets and expresses $\frac{4}{8} \times b$ u.	"$\frac{5}{8} \times 4$ ft $= \frac{5}{2}$ ft, because I broke 4 ft into eight equal parts, so one part is $\frac{1}{2}$ ft long. Then I took five copies of $\frac{1}{2}$ ft to make $\frac{5}{2}$ ft, or $2\frac{1}{2}$ ft." "$\frac{5}{8} \times 4$ ft $= \frac{20}{8}$ ft, because I made five copies of 4 ft and that is 20 ft. Then I split 20 ft into eight equal parts, so one of those parts is $\frac{20}{8}$ ft, or $2\frac{4}{8}$ ft." Teacher folds unit strip in half, then in half again, and then in half again. Student expresses teacher action as $\frac{1}{2} \times \frac{1}{2} \times \frac{1}{2} \times 1$ u $= \frac{1}{8}$ u. As teacher re-enacts slowly, student points to relevant portions of the expression.
Interprets and expresses $\frac{4}{4} \times b$ u as forming a product by split-copy.	"$\frac{5}{4} \times 4$ ft $= 5$ ft, because first I split the 4 ft by four, so one of the split parts is 1 ft, and then five copies of 1 ft is 5 ft." "$\frac{5}{4} \times 4$ ft $= 5$ ft, because five copies of 4 ft is 20 ft and one-fourth of 20 ft is 5 ft."
Expresses "half of half-unit" as multiplication $\frac{1}{2} \times \frac{1}{2}$ u $= \frac{1}{4}$ u. Interprets as split-copy.	Expresses "half of one-half foot" as $\frac{1}{2} \times \frac{1}{2}$ ft $= \frac{1}{4}$ ft. Writes $\frac{1}{2} \times \frac{1}{2} \times 1$ ft $= \frac{1}{4}$ ft, and interprets, "First you find $\frac{1}{2} \times 1$ ft, and then $\frac{1}{2} \times \frac{1}{2}$ ft, so it is $\frac{1}{4}$ ft." Writes $\frac{1}{2} \times \frac{1}{2} \times 1$ ft $= \frac{1}{4}$ ft and says, "One-half splits the half-foot, so that is quarters, and you take one copy of the quarter."

a length extends the meaning of product to include the generation of area. Movement of an area through a length results in a similar transformation of referent, now from area to volume. These movements can include fractional quantities, as depicted by Figure 7.6 for area and by Figure 7.7 for volume.

Scalar multiples of area and volume can also be generated, and some of these can be fractional scalars. For example, when constituting an area, rescaling one of the lengths by $\frac{4}{3}$ generates an area $\frac{4}{3}$ times the area generated by sweeping the unscaled lengths.

Another context is thinking of dividing as remeasuring. In this case, the quotient is the scalar multiple that relates the measurements that represent dividend and divisor. For example, consider a magnitude of length with measure 6 u.

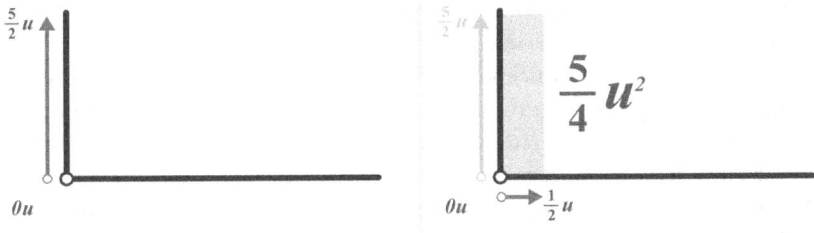

FIGURE 7.6 Sweeping fractional measures of length to generate fractional measures of area

If this magnitude is remeasured with a unit length one-third times as long, the result will be a measure 18 times as long as a one-third unit length: $\frac{6u}{\frac{1}{3}u} = 18$, and $18 \times \frac{1}{3}$ u is equivalent to 6 u. Quotients can also relate two fractional quantities of length (or of area, volume, etc.). For example, a fifth-grade student found the quotient of $\frac{\frac{3}{2}u}{\frac{1}{3}u}$ by first re-expressing each quantity in a common unit of measure, $\frac{\frac{9}{6}u}{\frac{2}{6}u}$, followed by imagining $\frac{2}{6}$ u as the unit of measure, so that four copies of this unit with another half of it appended would be just as long as $\frac{9}{6}$ u. The scalar multiple was then $\frac{9}{2}$, which she justified by symbolizing the measure as $\frac{9}{2} \times \frac{1}{3}$ u $= \frac{3}{2}$ u "because nine copies of $\frac{1}{3}$ u makes 3 u, and then the 2-split makes $\frac{3}{2}$u." Thinking of the quotient as a scalar multiple relies on a history of experience with multiplicative comparison and with understanding that the same magnitude of a quantity can have potentially multiple measurements.

We have suggested that measurement contexts open important opportunities for developing and extending students' understanding of rational number. Exposing students to these ideas from the early grades (in our experience, first and second grades) helps forestall the confusions that students sometimes experience later in elementary school when they run headlong into situations in which some of the principles of whole-number arithmetic seem no longer to apply. Partitioning and working with part units in contexts of measure can be extended from students' work with the same foundational concepts that underlie whole-number measure, such as unit iteration, iteration of composite

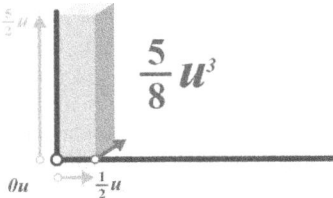

FIGURE 7.7 Sweeping fractional measures of area and length to generate fractional measures of volume

units, motion (e.g., length measure as a point along a path), the relations between measure and unit size, symbolizing measure, and additive and multiplicative comparison. Conceiving of fractions as measured quantities supports comparison of quantities grounded in operations of addition and subtraction. Framing fractions as operators grounds multiplicative comparisons between lengths and leads toward the generalization of a reciprocal relation between A measured in B and B measured in A. Metaphors of enactment in this case include the dynamic generation of fractional areas and volumes, as well as the copy-split metaphor for constructing fractional quantities.

In earlier chapters, we explained how measure contexts can support children's enactment and symbolization of arithmetic operations. Our goal in this chapter is to explain how measure concepts can reach even further into the elementary curriculum to connect to core ideas about fractions and rational number. Instead of presenting whole-number arithmetic, fractions, and measurement as modular topics, exposing and capitalizing on their inherent linkages can enhance students' understanding within each of these domains and lead to a more integrated and coherent conception of early mathematics.

References

Barth, H., Baron, A., Spelke, E., & Carey, S. (2009). Children's multiplicative transformations of discrete and continuous quantities. *Journal of Experimental Child Psychology, 103*(4), 441–454.

Beck, J. (2015). Analogue magnitude representations: A philosophical introduction. *The British Journal for the Philosophy of Science, 66*(4), 829–855.

Confrey, J., Maloney, A. P., Nguyen, K. H., & Rupp, A. A. (2014). Equipartitioning: A foundation for rational number reasoning. In A. P. Maloney, J. Confrey, & K. H. Nguyen (Eds.), *Learning over time. Learning trajectories for mathematics education* (pp. 61–96). Charlotte, NC: Information Age Publishing.

Confrey, J., & Smith, E. (1995). Splitting, covariation and their role in the development of exponential functions. *Journal for Research in Mathematics Education, 28*, 66–86.

Cortina, J. L., Višňovská, J., & Zúñiga, C. (2014). Unit fractions in the context of proportionality: Supporting students' reasoning about the inverse order relationship. *Mathematics Education Research Journal, 26*(1), 79–99.

Davydov, V. V., & Tsvetkovich, Z. H. (1991). The object sources of the concept of fractions. In L. P. Steffe (Ed.), *Psychological abilities of primary school children in learning mathematics* (pp. 86–147). Reston, VA: National Council of Teachers of Mathematics.

Depaepe, F., Torbeyns, J., Vermeersch, N., Janssens, D., Janssen, R., Kelchtermans, G. ... Van Dooren, W. (2015). Teachers' content and pedagogical content knowledge on rational numbers: A comparison of prospective elementary and lower secondary school teachers. *Teaching and Teacher Education, 47*, 82–92.

Lehrer, R., & Pfaff, E. (2011). Designing a learning ecology to support the development of rational number: Blending motion and unit partitioning of length measures. In D. Y. Dai (Ed.), *Design research on learning and thinking in educational settings: Enhancing intellectual growth and functioning* (pp. 131–160). New York, NY: Routledge.

Lehrer, R., & Slovin, H. (2014). *Developing essential understanding of geometry and measurement in grades 3–5*. Reston, VA: National Council of Teachers of Mathematics.

Ni, Y., & Zhou, Y.-D. (2005). Teaching and learning fraction and rational numbers: The origins and implications of whole number bias. *Educational Psychologist, 40*(1), 27–52.

Siegler, R. S., & Lortie-Forgues, H. (2017). Hard lessons: Why rational number arithmetic is so difficult for so many people. *Current Directions in Psychological Science, 26*(4), 346–351.

Smith, J. P., III (2002). The development of students' knowledge of fractions and ratios. In L. Bonnie & G. Bright (Eds.), *Making sense of fractions, ratios, and proportions. 2002 yearbook* (pp. 3–17). Reston, VA: National Council of Teachers of Mathematics.

Smith, J. P., III, & Thompson, P. W. (2007). Quantitative reasoning and the development of algebraic reasoning. In J. J. Kaput, D. W. Carraher, & M. L. Blanton (Eds.), *Algebra in the early grades* (pp. 95–132). New York, NY: Erlbaum.

Steffe, L. P., & Olive, J. (2010). *Children's fractional knowledge*. New York, NY: Springer.

Thompson, P. W. (1993). Quantitative reasoning, complexity, and additive structures. *Educational Studies in Mathematics, 25*, 165–208.

Thompson, P. W. (2011). Quantitative reasoning and mathematical modeling. In L. L. Hatfield, S. Chamberlain, & S. Belbase (Eds.), *New perspectives and directions for collaborative research in mathematics education*. WISDOMe Monographs (Vol. 1, pp. 33–57). Laramie, WY: University of Wyoming.

Thompson, P. W., & Carlson, M. P. (2017). Variation, covariation, and functions: Foundational ways of thinking mathematically. In J. Cai (Ed.), *Compendium for research in mathematics education* (pp. 421–456). Reston VA: National Council of Teachers of Mathematics.

Thompson, P. W., & Saldanha, L. A. (2003). Fractions and multiplicative reasoning. In J. Kilpatrick, G. Martin, & D. Schifter (Eds.), *A research companion to principles and standards for school mathematics* (pp. 95–114). Reston, VA: National Council of Teachers of Mathematics.

8
HIGHLIGHTS OF STUDENT LEARNING RESEARCH

We conducted research about student learning of concepts and practices of measure over the duration of the project (research on teacher learning is addressed in Chapter 9). As we will describe, the purpose, focus, and genre of research changed somewhat as we moved from Mallard Elementary School, where the development phase of the project was concentrated, to Sleeve Elementary School, where the tryout phase occurred. Nonetheless, at both sites, the overall purpose was to generate a portrait of the learning that emerged when students participated in instruction that was systematically responsive to potential trajectories of development, as envisioned by the learning progression outlined in Chapters 3–7. Indeed, the learning progression was itself an object of continuous improvement; its development both required and reflected changes in our research questions and methods over the span of the project.

The Two Phases of Research

There were two phases of research, although they overlapped somewhat. The first phase was conducted primarily at Mallard Elementary School and was designed to refine conjectures about the development of student learning, to inform curriculum development, and to guide emerging structures for professional development. This phase consisted of ongoing classroom observations of instruction; design research studies focusing on a focal subset of student conceptions (e.g., Brady & Lehrer, 2021; Wongkamalasai, 2018); and yearly, task-based, flexible interviews that allowed us to probe some of the nuances of students' ideas and practices. To illustrate the first phase of the research, we provide a broad review of students' responses to tasks posed during the flexible interviews in the realms of length, area, angle, and volume (one example of

DOI: 10.4324/9781003287476-8

the design research studies is presented in Chapter 2, and a second example is described briefly later in this chapter). Although it is not feasible within the confines of the chapter to undertake a comprehensive digest of the interview data (researchers conducted a total of 326 interviews of an average $1\frac{1}{2}$-hour duration), we report selected findings to exemplify both the student learning achievements and challenges that the data reveal.

The second phase began as the research transitioned to teachers and students at Sleeve Elementary. This phase, conducted in collaboration with colleagues at the University of California's Berkeley Evaluation and Assessment Research (BEAR) Center, addressed a question of contemporary interest within the assessment world—that is, whether and how teachers' ongoing, classroom-based judgments about student learning could be systematically integrated with data from summative measures of learning (i.e., in pre- and post-instruction tests) to generate more robust and comprehensive estimates of growth in learning. Integrating teachers' classroom data with traditional summative assessment data was a feasible enterprise only because teachers' judgments were focused on and calibrated to the same constructs as guided the development of the summative assessments. Moreover, as described in Chapter 2, the increased frequency and density of data provided by teachers contributed to the validity of an assessment system that otherwise depended primarily on broad-scale tests administered yearly. Another potential advantage of this kind of multi-tiered assessment system was that teachers' judgments about student learning would, on the one hand, contribute to psychometric modeling of student learning and, at the same time, provide ongoing streams of feedback to inform teachers' classroom decisions and professional practice (Chapter 9 presents a more complete description of the entailments of this kind of feedback for teacher professional development).

Constructing this multi-tiered assessment system required, first, the development of summative assessments administered at the beginning and end of each school year across grades 1–5 (kindergarten children were not included in these assessments, as many of them were not yet able to read the instructions). Summative assessments were constructed from a bank of 250 items and were tailored to the instructional goals pursued at each of the participating grades to provide a summary snapshot of progress indexed to the cognitive milestones (i.e., constructs) of the learning progression (Wilson, 2005). The items used in these yearly assessments were refined through multiple rounds of administration and revision at both Mallard and Sleeve.

The second and complementary constituent of the assessment system consisted of two types of teacher judgment about the nature of student thinking, both anchored to the conceptual milestones articulated in the strands of the learning progression described in Chapters 3–7. One format of teacher judgment was framed by teachers' choices of episodes of classroom instruction that they considered revealing of student learning. As Chapter 2 describes, teachers used TOTs (Teacher Observation Tools) to record episodes of class conversation, student work, and other evidence of learning that they felt indicated a conceptual performance by one or more students at a particular sublevel of the learning progression (Chapter 9 reports further detail about how teachers used TOTs and presents results of audit analyses of the evidence teachers generated to support their judgments about student learning). During any single class, a teacher observed a sample of students, but over time, all students could be (and typically were) observed about some aspect of their learning. During monthly teacher meetings, norms of evidence were negotiated as teachers shared their judgments of episodes of student learning.

The other format of teacher judgment consisted of teacher scoring of student responses to regularly administered formative assessments. Scoring rubrics were keyed to the conceptual milestones of the learning progression, as articulated in the constructs. The formative assessments, which were usually administered at the conclusion of each of the curriculum units, allowed teachers to collect a complete sample of student learning at a moment in time but, as compared with in-class observations of student learning, at the cost of a reduction in the teachers' ability to observe nuances of student thinking directly. However, follow-up formative assessment conversations that teachers conducted afterward to discuss the formative assessment results with their students allowed teachers to further probe the grounds of student thinking. Teacher judgments of student learning that were based on classroom observations and student responses to formative assessments were recorded in TOTs. Teachers used these records and accompanying visualizations to generate instruction responsive to student learning. The second major section of this chapter illustrates some of the forms of evidence of student learning generated by this attempt to design and develop an instructionally useful, psychometrically modeled assessment system.

Phase I: Student Conceptions of Measure as Indicated by Yearly Interviews

To inform and retune curriculum, interviews were initiated with first and second graders at Mallard Elementary, the first group of participants to enter the project in 2013. Subsequently, as instructional development for the higher elementary grades proceeded, kindergartners and third, fourth, and fifth graders were added. Accordingly, from 2013 to 2015, interview tasks focused primarily on length measure, with fewer tasks in angle and in area. When participating children began to transition to the upper elementary grades in 2016, both classroom instruction and the yearly interviews were expanded to include conceptions of area and angle more completely and also to include tasks designed to probe conceptions of volume measure. Altogether, there were four yearly waves of task-based interview data for length measure, three for angle, three

TABLE 8.1 Numbers of Students Who Participated in Task-Based Interviews Each Year

	2013	2104	2015	2016	Total students by grade
Kindergarten	0	21	20	20	61
First Grade	30	20	20	20	90
Second Grade	32	20	18	17	87
Third Grade	0	17	20	9	46
Fourth Grade	0	0	0	19	19
Fifth Grade	0	0	0	23	23
Total Students by Year	62	78	78	108	326

for area, and one for volume. As Table 8.1 shows, across the four years of interview administration, a total of 326 individual interviews were conducted with students from kindergarten through grade five. As the table reflects, the project began with first and second graders in 2013. In 2014 we added kindergarten and third-grade students. Fourth- and fifth-grade students were not included until 2016, reflecting the time that elapsed as students from the earlier grades progressed into the upper grades. Because additional participants were added in the early grades each year as the project proceeded, the total sample representing the primary grades is larger than the sample from the upper elementary grades. Mallard Elementary School's highly mobile population precluded a full longitudinal design, but we were nonetheless able to retain a modest number of students (70, or a little over a fifth of the total sample) who were interviewed across more than a single year. Altogether, 46 students were interviewed on two of these yearly occasions and another 24 on three occasions.

The kindergarten interviews took about half an hour to complete and were entirely performance-based, whereas interviews for children in the later grades varied from 60 to 90 minutes and featured items with a wider range of response modes (e.g., written multiple choice, computation, and short-answer questions; drawing; performance-based tasks with concrete materials). All interviews were video-recorded, and the student's notes, drawings, and calculations made as they worked were retained. The video recordings and student workbooks were used as data by researchers, who subsequently identified and scored the strategies that students employed to complete each problem (many of the tasks included multiple questions, each scored separately). At least two researchers independently scored each question; inter-rater reliability, calculated on each item, ranged from 84% to 91%. Disagreements were discussed and resolved by the original scorers.

Early-Developing Conceptions of the Measure of Length

We next describe a sample of the interview problems posed to probe early-developing conceptions of length measure, ranging from measure as direct comparison to equipartitioning and multiplicative comparison. Our focus in scoring was on identifying the nature of strategies deployed by children in kindergarten, first grade, and second grade as they responded to the problems posed. Figure 3.3 in Chapter 3 provides a visual summary of the relevant concepts and their relations that were targeted in the tasks and findings that we next describe as examples.

Direct comparison

In 2014 and again in 2015, kindergarten students were introduced to measure as comparison through a classroom investigation in which they directly compared paper-strip representations of the girths and heights of a set of pumpkins, usually at the onset of the year (as described in curriculum unit 1, length). At the end of the school year, we administered two performance-based interview tasks, both focused on the measure of length. The interviewer began by showing the child a cylindrical plastic can. Although the circumference was farther around than the can was tall, the perceptual illusion conveyed was that the height was greater than the circumference. The interviewer explained that the visiting researchers disagreed about which was greater, the can's height or its circumference. She provided a roll of 2.5-inch-wide paper tape and a pair of scissors and asked if these tools could be used to resolve the disagreement. The interviewer recorded what children subsequently did with the tools, along with their conclusions and justifications.

Upon first seeing the can, most children (in both 2014 and 2015) expressed the opinion that its height was greater than its circumference, a response that we expected. To compare the lengths directly, about two-thirds of the children cut the paper tape and used it to compare height and circumference. However, they did so in two different ways. Some children cut a single piece of tape to represent the height and then tried to wrap that tape around the circumference. When they found that it did not extend all the way around, they concluded that height was less than circumference. Others cut a strip to represent each dimension and then compared the strips. As they placed the strip against the can to assess its height, most kindergartners were careful to avoid observable gaps or overlaps between ends of the tape and the top and bottom of the can. However, when it came to placing the tape around the circumference, a small group failed to stop after circumnavigating the can and instead continued to loop the tape around the can, overlapping the ends of the tape. It is unclear whether children did this because they did not have a stable concept of the circumference as a length, because they did not know when to stop executing the "round and round" wrapping motion, or both. At both yearly administrations, another third of the children failed to create paper strips to represent dimensions of the can. Instead, they wrote numbers on a length of tape to generate a facsimile of a ruler and then used the "ruler" to "measure" first height and then circumference. However, they paid no attention to spacing those numbers equally on the tape, confirming our impression that although they believed that numbers were essential somehow for measuring, they did not conceive of numbers as indexing equal interval units on a scale. These findings, which were relatively stable over both years that this task was administered, indicate the accessibility of direct comparison of different attributes of an object mediated by representations of these attributes and also identify some of the challenges posed by this form of mediation.

Tiling, unit, and iteration

Kindergarten, first-grade, and second-grade students in 2013 and 2014 used different-colored laminated paper units of varying lengths (one, two, and three inches) to measure the edge of a 10-inch block of wood and reported the results of the measurement so that a teacher who was not present could make a block "just as long." Enough units of each length were provided so that students could tile the entire length of the block with units of a single magnitude of length, although it was not possible to use the longest unit exclusively and at the same time achieve a whole-number measurement.

Most of the students tiled multiple units across the entire length of the wooden block, and almost all of them used two different magnitudes of unit to complete the task (most of these students began by tiling with the longest unit, but then switched to a smaller unit when the as-yet-untiled span was shorter than the longest unit). Students tended to distinguish the units in their reported measure, as in "2 reds and a green," although a substantial minority did not, instead reporting a count, such as "3," perhaps because they knew that the interviewer could see which units were tiled or perhaps because they did not appreciate the role of unit labels in the reproduction of a measure. A couple of the kindergartners and first graders scooted a single unit forward while counting by employing a flawed form of iteration. These students moved the unit over arbitrary distances as they counted and failed to coordinate the motion of the unit with the count. Nonetheless, by the second year (2015) very few students, even in kindergarten, exhibited strategies other than tiling the length with units. Given these performances at task ceiling, we subsequently posed tasks to first- and second-grade students that elicited student critique and justification for properties of units, such as tiling and unit label.

For example, students were asked to evaluate faulty measurement strategies that were demonstrated by the interviewer and described as procedures endorsed by a hypothetical other child. The first portion of this task was aimed at assessing the children's awareness that when one measures a length, it is important to report the units that are used so that the measurement can be reproduced. The interviewer introduced the earlier-described plastic-laminated measuring units of lengths one, two, and three inches. Then she displayed a 6-inch line printed in the student notebook and explained, "Here is a line. John wants his friend, who is in a different room, to draw a line just as long in his own notebook. So, John measured, like this, and said that the line was 4 units long." At this point the interviewer carefully aligned the following units along "John's line": two blue 1-inch units followed by two red 2-inch units. The interviewer continued, "If John tells his friend that he should draw a line that is 4 units long, and if John has units like these, can he be sure his friend will draw a line exactly as long as this one?" We expected youngsters to say that John's friend would not know how to draw the line because John did not report the unit labels he had used to create his measure.

About 66% of the first graders appropriately objected that John could not be sure his friend would be able to copy the line because it was unclear which units John had used. However, the remaining 33% readily agreed that John, knowing only that it was 4 units long, would be able to re-create the line. Moreover, over the four years of data collection on this task, we observed little improvement among the first graders. This was probably due at least in part to first graders' difficulties with imagining the mental states of others. Young children often find it difficult to suspend what they know (they can readily see the line and the units used to measure it) in order to imagine what John's friend would know and think. (This finding inspired a subsequent change in instruction at Sleeve Elementary to include situations in which first-grade students attempted to reproduce measurements made by peers.) In contrast, second graders showed strong improvement on this item from the first year to the second. By the second year of data collection, 95% of them were objecting to John's use of mixed units, and 80% further explained why a measure of "4 units" was insufficient to guarantee that John's friend would get the same result.

After asking about John's strategy, the interviewer next demonstrated José's strategy for measuring a new (4-inch) line. The researcher demonstrated José's strategy by lining up five of the blue units but placing the units so that they clearly overlapped. During the initial year of the study, nearly half of the first graders accepted José's claim that the line was 5 units long. However, by the second year, all children in both the first and second grades objected to this procedure, explaining that if the units overlap, parts of the line will be measured more than once, resulting in a measure that is unduly large.

The interviewer then demonstrated Javon's strategy, which left noticeable gaps between the units. Over a third of the first and second graders accepted this measurement procedure during the initial year of the project, but the following year, after teachers implemented instruction in which children had opportunities to tile lengths and consider how gaps and overlaps influenced measure, only one child from each grade agreed with Javon's procedure for measuring. The remaining children pointed out that Javon was failing to measure "the gaps" and thus was under-measuring the length of the line.

Finally, the interviewer presented a new (pink, $\frac{1}{2}$-inch) unit and asked how the measure of Javon's line would change if Javon measured with pink units instead of blue. One third of the first graders struggled with this question when it was initially posed in 2013. However, by 2014, after students had received opportunities to experience measuring the same length with different units, it was challenging for neither the first nor the second graders. Some researchers (e.g., Sarama et al., 2021) have reported that it is especially difficult for youngsters to understand that a smaller unit yields a greater measure, but our interview data and classroom experience suggest that this idea is not inherently more difficult to understand than other concepts about accumulating units. It is likely that the apparent difference in findings is due to students finding it challenging to reason about this relation in the hypothetical situations that are often posed in assessments.

Unit iteration and symbolizations of unit on scale

First graders often resisted measuring a length when they had insufficient units to cover the expanse. For instance, in 2013, when asked to use a single unit to measure the side of a rectangle, 17% of the children insisted they could not do so because they did not have enough units. An additional 25% scooted the unit forward as they counted, but their counts were not aligned with the "scoots" and the distances traveled at each count were unrelated to the length of the unit. About 33% of the first graders attempted unit iteration but left gaps or overlaps between the end of one unit and the starting point for the next. Only 25% of the first graders completed the task successfully. Partly because the second graders did so much better (84% iterated the unit correctly and reported the correct measure), some of our teacher partners interpreted these findings as evidence of developmental constraints. Others decided to support unit iteration more explicitly by providing opportunities for first-grade children to explore its procedural (e.g., marking endpoints of intervals appropriately) and conceptual (e.g., coordinating count with unit translation) implications. The second wave of interviews a year later supported the teachers' expectation that these notions could be made sensible to the youngest students if instruction systematically supported experiences with unit iteration. By 2014, almost all the first and second graders (95% and 90%, respectively) iterated a unit appropriately to measure the length of a side of a rectangle. The remaining few children made minor errors of precision while translating the unit.

Finally, we assessed children's interpretation of units as equal lengths on a measurement scale and their understanding of the zero point as the origin of measure. Initially many children were confused about these ideas, but after a year of instruction we saw notable improvement. In one diagnostic task, the interviewer presented a tape measure with units demarked on it, but the units were not labeled, and children were asked to label them. In 2013, children showed a range of misunderstandings. About 33% of the first graders drew small lines all over the tape and added numbers at apparently random intervals (like the strategy the kindergartners applied to measuring the cylinder). They described what they were doing as "making a ruler," but their "rulers" did not feature equal partitions of units. In fact, there is no evidence that children thought of the lines that they drew as demarking partitions of length, because there was no discernible relation between the placement of the numbers and the lines, which were not equally spaced. Moreover, over 25% of the students labeled the zero point of the scale as "1," presumably because they knew that counting objects begins with 1. Only 14% of the first graders (but 47% of the second graders) labeled the origin as 0 and, in addition, labeled all whole units at the end of each unit, rather than in the center. However, as with iteration, by 2014 and following instruction focused on designing tape measures, children's performance on this task dramatically improved. By 2014, 90% of the first graders and 95% of the second graders correctly labeled all the whole-number units on the tape measure and also labeled the origin as 0.

Equipartitioning fractured units

Young children know that words like *half* refer to partitions but are not necessarily aware that the partitions are equal. To assess this understanding, the interviewer showed children a blue strip folded into two clearly unequal parts and said, "Julie told me that this [points to one of the strip's parts] is one-half of this whole blue unit. Do you agree?" During the initial year of data collection, a third of the first graders (who had not yet participated in instruction that included unit partitioning) happily endorsed Julie's suggestion that both sections of the strip were halves, a reply consistent with our observation that youngsters often use the term *a half* to refer to any piece. However, by 2014, all but one of the first-grade children objected that the strip was not folded in the middle, and therefore the two sections clearly were unequal and neither section could be considered half of the original unit. The second graders did not find this question difficult even during the initial year of interviews.

Measurement arithmetic

Students used multi-unit rulers to enact ("travel" on the ruler, as described in Chapter 3) and symbolize the addition and subtraction of whole numbers, halves, and fourths. Several of the interview tasks focused on these ideas of accumulating part units, an instance of part-unit iteration. For example, students were asked how many iterations of a half-unit would be required to travel from the origin to the end of an 8-unit ruler. Only a quarter of the first graders could solve this task during the first year's interview, and only half of those (about 13% of the sample) did so by mentally anticipating the number of iterations of the half-unit that would be required. The remaining successful children found the answer empirically by physically iterating the half-unit to the end of the ruler. In contrast, by 2014, 70% of the first graders correctly solved the problem by mentally anticipating, and another 20% solved it by iterating the half-unit to the end of the ruler. The second graders did even better; 75% of them successfully solved the problem in 2013 (60% by mentally anticipating the results of iteration). By 2014, 90% of the second graders solved the problem and all but two did so mentally—that is, without moving the half-unit.

In addition to the operations of addition and subtraction, we also sought to learn whether and how children were beginning to understand the multiplicative relations between units and their partitions. For instance, the interviewer showed a student a long strip of orange paper (referred to as an orange unit) and a small partition that was described as one-tenth of the original orange unit. The child was then asked, "How many of these smaller parts [referring to the $\frac{1}{10}$ unit] would I need to put together to make a strip just as long as one whole orange unit?" This item was first introduced during the second year of data collection. At that time, about half of both the first and second graders understood the logic that partitioning a strip into 10 congruent pieces means that one would need 10 partitions of size $\frac{1}{10}$ strip to restore the original unit. The interview included several other tasks designed to assess students' multiplicative reasoning. In general, children's multiplicative reasoning was stronger when the relation between two lengths was expressed as a whole number. For example, nearly half of the second graders were able to draw a line "three times as long" as a model line shown in their notebook. In contrast, only 15% were able to draw a line "one-third times as long" as a standard line. Multiplicative comparisons were not understood as reversible even by older students. Three-quarters of the third graders successfully filled in the blank in the sentence "3 units is _____ times as long as $\frac{1}{2}$ unit." In contrast, only 15% of them gave the correct answer to "$\frac{1}{2}$ unit is _____ times as long as 2 units." Findings like these influenced us to recalibrate professional development efforts with educators to focus on describing reciprocal multiplicative relationships between units and the lengths they are used to measure. That is, a ruler is 4 times as long as 1 unit, *and* the unit is $\frac{1}{4}$ times as long as the ruler. In later years at Mallard, children's performance improved on items like these, but many students continued to struggle with the idea that a multiplicative relationship could be described as a fraction. Children's explanations suggested that thinking this way conflicted with their expectation, based on earlier instruction, that multiplying is associated with thinking about "groups of" and necessarily results in an increase in the original value.

Conceptions of Area Measure

As our work with teachers continued, we gradually extended the reach of instruction beyond the primary grades. Next, we highlight problems posed to students in grades 3–5 to reveal their conceptions of area and its measure, with an emphasis on unit structuring, unit and other dissections of area, differentiating length (e.g., perimeter of figures) from area measure, and dynamic generation of area. Interviews conducted in 2014 reflected student conceptions of area and its measure with traditional instruction, whereas those conducted in 2015 and afterward reflected conceptions of area developed with explicit support for unit and for area as dynamically constructed by sweeping lengths (as described in Area curriculum units 1–3). Dynamic conceptions of area were examined more closely in design studies conducted at the third-grade level in 2015 and 2016. In the 2016 study, students employed physical and digital tools to conduct investigations of the generation and measure of area, and a sample of participants were interviewed to probe their conceptions of fractional unit areas and dissections of parallelograms into rectangles.

Necessary conditions for area

Third-grade students examined several two-dimensional shapes and were asked to circle those that "have an area." As Figure 8.1 shows, the 2014 and 2015 versions of this task differ somewhat.

Both versions of the task include some shapes that are unfamiliar, such as shapes A, B, and C in 2014 and shapes B and D in 2015. Unfamiliar shapes were included to learn whether students assume that the only shapes that have area are familiar two-dimensional figures encountered in math class. In 2015, the task was modified to include Figures A and F, to learn how students resolved two debates that emerged that year during instruction in class: (1) whether figures that are not closed can be said to have area and (2) whether a line should be interpreted literally as an object on paper—which necessarily has an area—or is intended to represent a one-dimensional length.

In 2014, only a single third grader of the 17 interviewed correctly claimed that all the shapes have an area. Most (65%) selected only C, D, and/or E, suggesting that familiar geometric shapes encountered in previous mathematics instruction were the only figures considered to have area. An additional five children (29%) included either A or B, unfamiliar figures, in their choices. In contrast, in 2015, 90% of the third-grade students nominated all closed figures, ruled out A and F (the line and the open figure, respectively), and, in addition, provided plausible justifications for their choices. These include, for example, explaining that because F is not closed, its area is indiscernible because a prospective count of units has no end, or that A was not chosen because although the literal illustration of a line has area, a line conventionally represents linear distance, which does not have area.

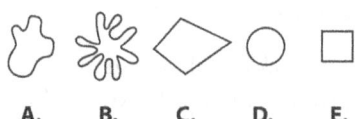

     A.   **B.**   **C.**   **D.**   **E.**

FIGURE 8.1a Students were asked to circle the shapes that "have an area," 2014 version

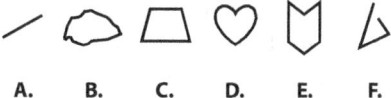

FIGURE 8.1b   Students were asked to circle the shapes that "have an area," 2015 version

Unit structuring of area

To reveal how students coordinated intervals of length measure to structure a figure as an array of square units, third-grade students determined the area of a rectangle with labeled sides of 5 in and 3 in and drew an approximation of the resulting array of square units. We looked for evidence that students attempted to construct equal intervals for each side length and coordinated these units of length to dissect the area into a grid of square units. In the 2014 interview, only one student (of 17) could do so. In the 2015 interview, 75% of the 20 third-grade students constructed an array of 15 square units. For these students, the interviewer next posed a more challenging variation of the task in which the units of measure were not identical, with side lengths of 2 K and 3 T. Seventy-five percent of those students who tried this problem coordinated the nonidentical units of length measure to dissect the figure into six identical rectangular units. Eighty-one percent of those students reported an answer of six units, although only 65% correctly labeled the unit as TK.

In the following year (2016), although 89% of fourth-grade students (total $n = 19$) computed the measure of the area of a 3 u × 4 u rectangle, fewer (79%) were able to dissect the rectangle into square units by mentally coordinating intervals of length measure. To probe students' capacities to employ dissection strategies more generally, we asked fourth-grade students to find the area of the concave polygon displayed in Figure 8.2, in which unlabeled sides were

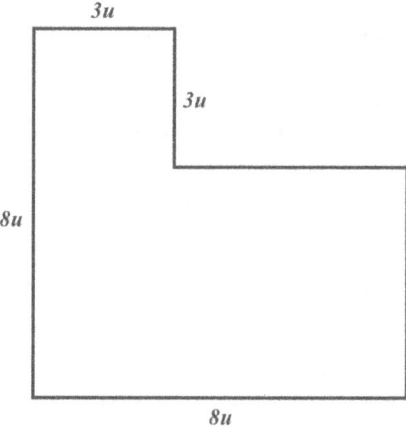

FIGURE 8.2   Finding the area of a polygon with two unlabeled sides

not given. We anticipated that students would employ known properties of rectangles to find the missing length and a dissection strategy to obtain a measure of the area. For example, students could find the area by visualizing the area measure of an 8 u square and then subtracting the area measure of a 3 u × 5 u rectangle. Or, students could dissect the figure into a 3 u × 3 u square and 5 u × 8 u rectangle.

Students used a range of dissection strategies to find the area; of these, over half were correct. For example, 26% of the students found the area measure of an 8 u × 8 u square and then subtracted the area of a 3 u × 5 u rectangle. Sixteen percent (16%) added the area measure of the 3 u × 3 u square to the area of the 5 u × 8 u rectangle. Two students, or 10.5%, added the area of a 3 u × 8 u rectangle to the area of a 5 u × 5 u square. However, 42% of the students could not develop an appropriate dissection. Most of these students simply multiplied the lengths of the two longest sides (that is, 8 u × 8 u) and reported an answer of either 64 units or 64 u².

Differentiating area and length measure conceptually and symbolically

As noted, fourth-grade students were shown a rectangle with dimensions labeled 3 u and 4 u. Nearly all students (90%) differentiated area from perimeter measure and determined the measure of the rectangle's perimeter. However, consistently symbolizing distinctions between length and area proved more challenging. For example, fourth-grade students were asked to distinguish between 2 × 2 ft and 2 ft × 2 ft. Most (63%) focused solely on the magnitude of the product and claimed that the expressions "mean the same thing." Fifth-grade students were more attuned to this distinction, with 82% attributing either an iterated length or a collection of lengths to the first expression and an area to the second expression. An additional 9% of fifth-grade students noticed the difference in unit but could not draw a representation of the difference.

When students were asked to represent $\frac{1}{2}$ u × $\frac{1}{2}$ u, 48% of fourth-grade students drew a $\frac{1}{4}$ square unit, but only three of the students (16%) labeled it as $\frac{1}{4}$ u². The remaining students (32%) labeled it $\frac{1}{4}$ u. In contrast, 73% of the fifth-grade students drew a $\frac{1}{4}$ square unit, although again, fewer than half (46%) appropriately labeled it as $\frac{1}{4}$ u². The remaining 27% labeled it $\frac{1}{4}$ u.

Looking across tasks and years, we concluded that students tended to conceptualize area as space enclosed and, increasingly, as space generated by moving a length through a second length. Most students could envision side lengths of rectangles as intervals of length measure and could coordinate these intervals to dissect rectangles into square units. It was more challenging to structure a rectangle as an array of units when the measures of the lengths of the sides were not the same, but nonetheless, a substantial percentage of the third-grade sample did so.

Students readily distinguished between length (perimeter) and area measures of the same figure. However, student responses to tasks that probed their interpretations of symbolic expressions revealed a bias toward focusing on magnitude, so that expressions with the same numeric value were conceived of as equivalent. Similarly, unit labels appeared to be considered optional and not critical to area measure. This result suggested a design failure that apparently produced parallel worlds of symbolic expression and tangible manipulations that generated areas or that dissected areas into units. We subsequently revised instructional tasks and critical forms of teacher support to provoke greater integration between mathematical activity and its symbolization.

For instance, in a follow-up design study conducted the following year with third-grade students (Brady & Lehrer, 2021), we amplified support for area measure by developing a digital tool that allowed students to investigate dynamic generation of area and to conduct dissections by units or by cuts of their choice on figures generated dynamically. At the conclusion of instruction, we interviewed a representative sample ($n = 8$) of the third-grade student participants. Presented with a figure depicting a 2 in × 4 in rectangle generated by sweeping the short side through the long side, students determined the measure of its area and perimeter and coordinated intervals of length to dissect the rectangle into square units drawn on the figure. All students differentiated between measures of perimeter and area, and all but one dissected the rectangle into inch-square units by coordinating estimates of inch intervals of length. Students also considered a rectangle generated by sweeping a length with measure 3 T through another length with measure 2 K, where K was not congruent with T. All but one student indicated that the measure of its area was six rectangles generated by coordinating unit intervals of different lengths (T, K). To measure perimeter, students aggregated lengths of T and K, but only two students proposed 6 T + 4 K as its measure; the others tended to focus on the count, as in 10 T & Ks.

We also posed problems in area measure with access to the digital tool where intervals of measure were visible on horizontal and vertical scales. Challenged to generate and determine the area of a rectangle with one side length having a fractional measure (3 in × $4\frac{1}{2}$ in), seven of the eight students determined the measure of its area by visualizing composite units, usually either as three rows of four square units or as four columns of three square units, and then visualizing the composition of the remaining half-square units to constitute a measure of $13\frac{1}{2}$ square inches. Students observed the formation of a parallelogram made by sweeping a length at an angle other than 90°, with vertical height of 5 in, through 8 in oriented horizontally. Students determined its area by visualizing or by using the digital tool to dissect the parallelogram into a rectangle. Then they visualized composite units (e.g., "five rows of eight squares or eight columns of five squares") to obtain the measure of its area. Finally, we asked five students to interpret the result of movement of $\frac{1}{2}$ u through $\frac{1}{2}$ u when the lengths were oriented perpendicularly. Three students immediately replied that the motion generated $\frac{1}{4}$ u². The other two students suggested "a square $\frac{1}{2}$ in" which was "one-fourth of a big square inch" and "one-fourth of that whole thing," gesturing and tracing a u × u square on the screen.

Conceptions of Angle Measure

Most classroom instruction on angles and their measure depicts angles as static figures, and the measure of an angle is commonly described as depending on the amount of opening between its sides. However, thinking about amount of opening can be confusing for children who may conceive of angles as literal objects and who therefore do not understand that the lengths of the angle sides are arbitrary and do not affect the angle's measure (many children expect that an angle drawn with longer segments is greater than the same angle portrayed with shorter segments). To counter this confusion, we sought to help children integrate more traditional conceptions of angle-as-figure with angle-as-turn about a fixed point. The turning metaphor is consistent with our general design emphasis on capitalizing on children's intuitions about bodily motion. Teachers carried out some initial tryouts of lessons on angle in the first and second grades but were reluctant to focus much instructional time and energy on angle measure, which is a topic that local and state standards framed as appropriate for older students. Hence, young children's opportunities to conceptualize angle as turn were comparatively sparse.

To coordinate the angle-as-figure perspective with the angle-as-turn perspective, students used stakes and ribbons to construct rectangular walking paths on the school's lawn. Students first generated and then enacted instructions that others could follow to make a copy of a path. As explored in decades of earlier research conducted with Logo (Papert, 1980), walking a path entailed differentiating movement to generate a length (and units of length magnitude) from the turning movement (including axis of turn and units of magnitude) necessary to alter direction of travel. During classroom instruction, enactments of walking paths included animating a path route taken by a small Lego™ figure as it "walked" on a whiteboard, to complement the paths that children literally walked in large-scale space. During instruction, teachers and children found it necessary to distinguish between exterior and interior angles, so the former were designated as *turn angles*, and a bodily metric of one complete rotation, a whole turn, with increments of $\frac{1}{2}$ wt and $\frac{1}{4}$ wt, was developed to measure magnitudes of rotation.

Embodied turns in walking paths in primary grades

In 2015, the interviews for first and second graders included a task to assess a path perspective of angle. Students were shown a rectangle drawn on paper with vertices labeled A, B, C, and D. They were given a 2-inch plastic unit to serve as a measure of length and a small Lego™ figure. The instructions were "Here is a starting point, A. I'd like you to write the directions that will tell someone how to draw this rectangle if they start at A and use this unit to describe the length of the path." The model of the path to be "walked" (e.g., the rectangle) guided students as they generated their directions. At the same time, the rectangle also provided a test of those directions (e.g., if my Lego-guy follows the rules that I have proposed, does he faithfully follow the model rectangle displayed on paper?).

In 2015, and again in 2016, first-grade students differentiated movements for length from those for turning, but only a quarter of the students correctly described the turns in the rectangle path as one-fourth, or one-quarter, turns to the right. However, the second graders did somewhat better; 56% in 2015 and 59% in 2016 correctly labeled both the direction and the amount of the turns. Most of the remaining students either neglected to include the direction of turn or indicated an incorrect direction on at least one vertex, typically the final turn, as the Lego™ figure turned back toward its initial heading. Many students described this as a left turn, presumably because it was toward *their* left, instead of noting that, from the perspective of the Lego™ figure, the turn was a turn to the right. Third graders also attempted this task in 2015, the first year they were included as participants in the yearly interviews. Their performance was much like that of the second graders; 56% of the third graders described the turn at each corner of the rectangle as a "one-quarter turn" to the right, and an additional 27% of them erred only in the direction of the final turn. (The following year, 2016, we focused primarily on third graders' conceptions of area and did not ask them about angle.)

Conceptions of angle and measure in later grades

Fourth-grade students ($n = 19$) also participated in formative interviews for the first time in 2016, the final year of interview administration. The fourth-grade interview featured a more extensive set of tasks for assessing students' understanding of angle measure, in light of the greater emphasis on angle measure prescribed by the state standards. Although much of the instruction focused on measure of angles as figures, teachers also introduced students to path perspectives on polygons and to the use of protractors, including circular protractors, as tools for measuring angles.

When asked what an angle means, 47% of the fourth-grade students adopted a view of angle-as-figure, describing an angle as "where two lines meet at a vertex." Two additional students described angle as part of a shape—e.g., the corner of a rectangle. In contrast, about 37% of the fourth graders described angle as a turn or amount of turn, consistent with the instructional emphasis of the angle lessons on a body-centered, movement-based concept of angle. The interviewer next reminded students that we measure angles in degrees.

"But what," she asked, "is meant by a degree?" Three students (16%) were unable to reply. Although 26% reported that a degree has something to do with how an angle is measured, they could not say anything more specific. An additional 31% said that a degree tells something about the turn of an angle but did not say that degrees are measures to quantify amount of turn. Finally, 21% defined a degree as $\frac{1}{360}$ of a whole turn. Despite their somewhat tenuous grasp on a definition of a degree, students' estimates of magnitudes of rotation expressed in degrees were usually accurate. Rotating a pencil about a point, most students correctly illustrated turns of 90° (95%), 180° (90%), 360° (90%), and 135° (79%).

The students also saw two figures of 45° angles, juxtaposed side by side. However, in one figure, the segments were considerably longer. Students were asked whether the two angles had the same or different measures. Although two students said the figure with the longer sides was a larger angle, 90% correctly stated that both angles had the same measure; and most of them (74% of the sample) said this was because the openings of the angles were the same, even though the sides were not the same length. Two students (10.5%) replied that both figures show the same amount of turn at the hinge. A final student did not justify his reply.

A few of the interview items assessed children's understanding of and procedural competency with measurement tools. For instance, students were asked to explain the meaning of the markings indicating degrees on a circular protractor and to clarify why there are two circles of degree measures on the protractor. A few (10.5%) of the students simply said that they did not know, and another 10.5% said that the markings indicate degrees. The remaining 79% said that the markings indicate degrees and that the two different circles indicate different directions of angle measure. The interviewer next displayed a 120° angle and asked students to show how to use the protractor to measure it. Although a little over half (53%) used the protractor correctly and reported the measure, 47% used it incorrectly.

Understandings of angle theorems

We also posed a series of questions about theorems, such as the supplementary angle relationship that arises from a line's half-turn symmetry and the 360° theorem that arises from considering a polygon as a path (the turn angles must sum to 360° if the walker returns to her original heading after completing the path). Students first inspected a drawing of a horizontal line intersected at a 45° angle by another straight line. The 45° angle was labeled, and students were asked about the measure of the supplementary angle. Sixty-three percent identified the measure of 135° by subtracting 45° from 180°. An additional 10% found the measure by using a protractor, indicating that they did not see the necessity of the relationship. A quarter of the students produced an incorrect answer. Most of them tried to estimate the measure of the angle visually, although one tried to justify an answer obtained by subtracting 45° from 360°.

Students were shown a figure of an equilateral triangle and told that all the sides were the same length. They were then asked to indicate the locations of the turn angles. Almost all the students correctly did so, and 69% also recalled that the sum of the turn angles would be 360° because that was the same as one whole turn, the extent of turning necessary to start and end a path facing in the same direction. When asked "How much is each turn angle?" over half the students answered correctly. When asked to explain their answers, those who were correct either found the turn angles by subtracting the measure of the interior angle from 180° (31.5% of the sample, or six students), by dividing the total number of degrees in a whole turn by the number of angles (21%, or four students), or by estimating the angle or using a protractor to measure it (10.5%, or two students). The remaining children (31.5%) used a range of strategies that produced incorrect answers, including incorrectly measuring with a protractor or misapplying a theorem or rule (one student did not answer this question). When asked about the measure of the "inside angles" of the equilateral triangle, 57% of the students correctly explained that each inside angle equals 180° minus the turn angle (that is, 120°), although 10.5% of the students who suggested this procedure could not explain why it works.

Students were next shown a figure of a right triangle and told: "Leona thinks that when you add up its turn angles, it will be 360°. But Linda disagrees. She thinks that because it is a different kind of triangle (that is, different from the previous equilateral triangle), the sum of the turn angles will be different. What do you think?" Seven students endorsed Leona's correct answer, and five students agreed with Linda (the remaining students either were not asked or did not answer this question). When asked to provide a justification, most of the students who were correct (five students, or 26% of the entire sample) pointed out that moving all around the triangle requires making one complete turn, or 360°. Two, or 10.5% of the 19 students, attempted to justify their reply by measuring each turn angle and summing the angles, and so these students assumed that the question required empirical verification. Two students simply said that the equilateral and right triangles are different, so the turn angles must differ.

Finally, students were asked about the sum of a regular hexagon's turn angles and the measure of each interior angle (five of the students were not asked this set of questions because they did not answer the previous question correctly). Eleven of the 14 children who responded (58% of the total sample) correctly said that the sum of the turn angles would be 360°, and all but two of these students pointed out that if each turn angle is 60°, the sum of six angles will be 360° (the remaining two attempted to find the measure of each turn angle and then sum them). Only 31.5% of the students, however, correctly explained that each interior angle of the hexagon would be 120°, or 180° minus the turn angle of 60°.

Conceptions of Volume Measure

Although conceptions of volume and its measure ideally codevelop with other realms of measure, state and district standards in play at Mallard Elementary designated fifth grade as the locus of instruction in volume measure. We designed instruction in volume measure to parallel the approach followed in area measure. That is, instruction included opportunities for students to investigate unit and composite-unit structuring of the volume of right rectangular prisms and to learn to conceive of dynamic generation of volume as a product of area and length. Dynamic generation set the stage to consider the volume measure of cylinders, right prisms with different bases (e.g., hexagon, triangle), and, through application of Cavalieri's principle, non-right prisms. (We describe some of this instruction more fully in the text that follows, which reports evidence of student learning generated at Sleeve Elementary.) In 2016, the yearly interview for fifth-grade students ($n = 22$) at Mallard Elementary focused on this range of students' conceptions of volume measure.

Strategies employed to measure prisms constructed of cubic units

The first two interview problems asked students to find the volume of prisms constructed with unit cubes. In one of the structures, not all the units were visible, although the lattice of units was marked and therefore visible on the outside of both of the prisms. In addition, students were shown a wooden cube with a volume of 1 in^3 and were told that this cube would serve as the unit of volume for the next couple of questions.

The first prism was a tower constructed of 2 in × 2 in × 5 in wooden cubes. Students were instructed to find its volume. Most (86.4%) of the 22 students correctly said that the volume was 20 cubic units, and an additional student used an appropriate strategy but made an arithmetic error. The remaining two students counted the faces of the cube and reported the total. Students applied three different appropriate strategies (although two students were unable to identify a strategy). Five students (23%) found the dimensions of the tower by referring to the edges of the cubes and then multiplied the dimensions. Nine (41%) counted the number of cubes in one layer and then counted by layers (e.g., by 4s). A final 27% counted each visible block, one by one.

Next, students found the volume of another tower with a hidden (e.g., not visible from any perspective) unit within the structure. Seventy-three percent of the students found the correct answer of 36 cubic units, and two more (9%) used a reasonable strategy (e.g., iterating a layer or composite of cubic units) but made an arithmetic mistake. Once again, 23% measured the dimensions and multiplied, but most (nearly 60%) counted the number of cubes in one layer and then counted by layers.

Strategies employed to measure prisms with partial structuring

Students next found the volume of an unmarked, hollow rectangular prism that was labeled 3 in × 4 in × 3 in. The faces of the prism were not inscribed to indicate unit cubes, but students had access to enough wooden cubes (1 in³) to partially structure the volume by, for instance, lining a row and column of the lattice to support visualization of the rest of the volume (e.g., as layers of composite units). There were, however, insufficient cubes to fill the entire box. Students also had access to grid paper and a ruler.

Fifty-five percent of the students found the correct answer of 36 cubic units. Three students (14%) used either the ruler or the blocks to measure each dimension of the prism and then multiplied to find the product. Eight (36%) found the area of the base and envisioned the volume as the product of the area times the height. Three (14%) arranged cubes to cover the base of the prism and attempted to count by layers, and another 14% tried to count with individual blocks. The remaining five students used inconsistent strategies or said the task was impossible.

Next students were given a rectangular prism that was 3 in × 4 in × 4.5 in, along with several wooden cubes (1 in³), but not enough to fill the box. They were also given paper and a ruler. As neither the box nor the cubes were marked, students had to mentally construct the half-units to solve this problem. As might be expected, performance was somewhat lower than on the immediately preceding item. Here 32% of the students found the correct answer of 54 cubic units, although an additional 32% used a reasonable strategy to measure (e.g., using tools appropriately to find the dimensions of the box) but made one or more arithmetic errors in calculating the volume.

Volume of pentagonal prism

The interviewer presented a rectangular, pentagonal right prism, with each base side of 3.75 in and a height of 4 in. The same tools were available, and students were asked to find its volume. Now 27% of the students found the correct answer of around 96 cubic inches, and an additional 19% used the tools appropriately to measure dimensions but made an arithmetic error. The students who succeeded approximated the area of the base with grid paper and then multiplied the area by the height. Three of those who failed continued to try to count with individual blocks. The remaining students who failed either could not devise a strategy or used a strategy whose underlying logic we could not discern.

Cavalieri's principle

Cavalieri's principle specifies that two structures with the same cross-sectional area and height occupy the same volume. The principle is another instance of a dynamic conception of volume as generated by the motion of an area through a height. Two items probed students' thinking about this principle. The interviewer first presented a right cylinder and another that was oblique. Students were shown (by demonstrating their congruence) that the bases of each cylinder had the same area and were told that the heights were also the same. Then the interviewer asked whether one cylinder had more volume than the other or whether the volumes were the same. Two-thirds of the students said the volumes were the same. A few (13.6%) explained by referring to a dynamic image—namely, that they could imagine rearranging thin layers of the oblique cylinder to produce a copy of the right cylinder. Others (45.5%) said that the oblique cylinder was the same as the right cylinder, only tilted. Next, the interviewer presented a stack of cards to form a right rectangular prism and handed the students a stack of the same number of cards. Students were asked to use the second set of cards to produce another shape with the same volume. Forty-one percent of the fifth graders used the cards to generate a three-dimensional stack tilted at an oblique angle. They justified their solutions by explaining that both stacks have the same base and the "same distance around" or by saying that both stacks "fill up" the same amount of volume. However, 45.5% of the students simply rotated the stack 45° or 90°, and another 13.5% were unable to produce a shape that they believed would have the same volume.

As the preceding examples demonstrate, the yearly interviews were critical for revealing aspects of the instruction that were effective, as well as for exposing weaknesses that needed to be addressed with further retuning of instruction. These interview results suggested that much of the instruction was working as expected, but the extensive probing format of the interviews made it impossible for us to conduct them with more than a small, representative sample of the participating children. As we have demonstrated, in many cases the justifications and subsequent probing by interviewers revealed ways of thinking that were important for both researchers and teachers to understand. Because of their richness and depth, data from these interviews guided ongoing revisions to curriculum and instruction, in addition to informing teachers about the nature and range of children's thinking about the mathematics of measure. Additional data about student learning were derived from the assessment results generated by all participating students at Sleeve

Elementary during the second phase of the research, after the curriculum and instructional approach had achieved more stability.

Phase II: Summative, Formative, and *In-Situ* Evidence of Student Learning

During the second phase of research, evidence of student learning was inferred from student responses to both summative and formative assessment items, as well as from student talk and activity as these emerged during the conduct of instruction at Sleeve Elementary. These different sources of information were integrated by employing constructs as a common interpretive framework and were further transformed into quantities with a common scale, with statistical models developed by our colleagues at the University of California, Berkeley, to estimate progress in student learning across these diverse contexts (see Lehrer &Wilson, 2021; Wilson, 2021, for descriptions of the modeling approach). In the remainder of this chapter, we illustrate how student responses in each of these distinctive contexts were employed to further articulate the learning progression.

Summative Assessment

Traditional summative assessments, such as those commonly employed for purposes of statewide accountability in the U.S., support inference about the status of student learning from student responses to items. Items are constructed to differentiate among students, but the relation between a student's response to an item and the particulars of the student's conceptual development are usually not well specified. Clearly, some students achieve more than others, but it is often difficult to employ student responses to summative items as productive guides to future instruction (Horn, Kane, & Wilson, 2015; Pellegrino, Chudowsky, & Glaser, 2001; Shepard, 2010).

In contrast, construct-centered item design rests on a conjecture about the relation between a student response and how the student is likely conceptualizing a particular idea or enacting a specific skill (Wilson, 2005). Constructs articulate the qualities that are quantified by a summative assessment. For example, when a student responds correctly to the first part of the item displayed in Figure 8.3, we infer that he or she conceptualizes a qualitative compensation between the magnitude of a unit and a measure and understands that a measure of a magnitude of a length in shorter units is greater than its measure with longer units. This way of thinking is described at the second level of the length construct. A correct response to the second part of the item further indicates that the student conceptualizes this relation of compensation multiplicatively, a way of thinking that is associated with the fourth level of the length construct. In this manner, every item in a summative assessment indicates one or more sublevels of a construct, and summative assessments are designed to sample different levels within a construct and, occasionally, multiple constructs. As noted earlier, we designed and tested items for each primary measurement construct (length, area, volume, angle) and, with our colleagues at the BEAR center, generated and administered summative assessments at grades one through five over the course of the project.

Summative assessment contributed to the articulation of the learning progression in several ways. First, teachers viewed student responses to individual items or clusters of items that were administered before instruction as guides to ideas they should emphasize during instruction. For example, teachers at early grades noticed that before systematic instruction, students typically did not conceptualize a qualitative compensation between the magnitude of a unit and a measure, as indicated by their responses to items like the first one in Figure 8.3. As a result, teachers deliberately highlighted student explanations of this relation during subsequent instruction. Second, pre-post changes in scaled scores on these assessments (see Wilson, 2005, for how student responses to a set of test items can be calibrated to share a common scale) provided estimates of the extent of conceptual change (in a log odds metric) in each construct across relevant grade levels. For example, in the third year of the project, student conceptual change on the length construct in the first grade exceeded 1 logit (i.e., 1.1), and similar extent of change on this construct was evident across grade levels. Statistically significant change was also evident for the area, volume, and angle constructs. Hence, summative assessments provided a snapshot of the extent of construct-centered progress. Summative estimates were positively associated with scaled estimates of student learning derived from teacher judgments, $r = .81$ (Wilson, 2021). Hence, teacher judgments of student learning were triangulated with more traditional summative assessment.

Third, because students' conceptions and item difficulties are represented on the same scale, relations between them can be examined in light of the image of development implied by a construct. Items indicating later-developing

---

Sam measured the length of a room with his feet, walking heel to toe. His father also measured the length of the room, walking heel to toe. His father's foot is longer than Sam's foot. One measure of the room was 18 *feet* and another measure was 12 *feet*.
Which measure was Sam's?
Sam's measure was:

○ 18 *feet*

○ 12 *feet*

Sam's father's foot is _____ times as long as Sam's.

○ $\frac{2}{3}$ ○ 2   ○ $\frac{1}{2}$ ○ $\frac{3}{2}$

FIGURE 8.3 Assessment item that focuses on comparative magnitudes of unit length and measured length

conceptions are anticipated to be more challenging (i.e., require greater conceptual development for success) than those indicating earlier-developing conceptions, and items indicating the same level of performance on a construct ideally should be equally difficult (see Lehrer, Kim, Ayers, & Wilson, 2014, and Wilson & Lehrer, 2021, for a fuller description of the approach employed). Of course, these ideals are subject to disruptions, such as potential effects of item response type (e.g., multiple choice vs. constructed response) and the intelligibility of situations posed in an item, such as that depicted in Figure 8.3, so patterns of item difficulty are always more heuristic than rigorous test.

Figure 8.4 illustrates pre-post change in conceptions of area measure during the third year of the project (a year in which instruction was a mix of online and in-person formats). The left side of the figure indicates the distribution of estimates of individual students, the middle portion shows item difficulties, and the right-side bands correspond to levels of the construct. This figure suggests that, as anticipated, students usually found items indicating later-developing conceptions of area measure to be more challenging than those indicating earlier-developing conceptions. For example, sweeping fractional measures of length, an item at the fourth level of the area construct (i.e., dynamic conceptions of area and product), is more difficult than an item indicating assembly of fractional units to measure an area, an earlier-developing conception. However, there are also some surprises, such as in distinguishing the area and the perimeter of rectangles when the measures of the sides are not in the same unit, as perhaps foretold by the interview results of the design study in area measure described previously. Students responding to items indicating this distinction often dissected area by coordinating different unit intervals to constitute appropriate rectangular units of area measure (e.g., a t × s unit). However, although they summed the values of the lengths of the sides, some of them failed to preserve the distinction between units (e.g., 14 instead of 6 t + 8 s). Thus, student responses clearly distinguished between perimeter as a length and area as an enclosed space, but fewer than expected preserved

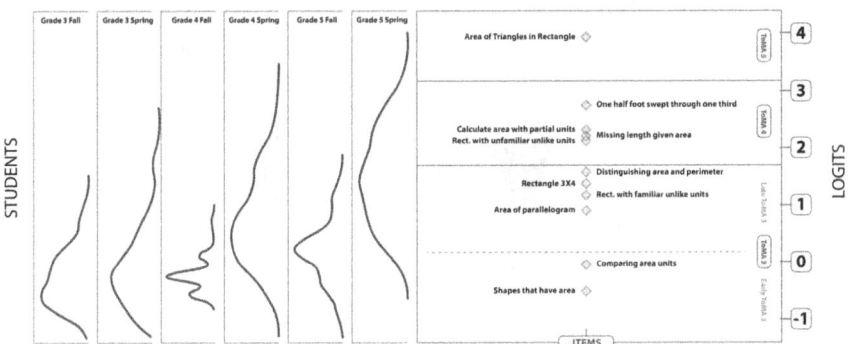

FIGURE 8.4 Pre-post change in conceptions of area measure

this distinction in their choice of unit of measure. Note, too, that the bands of items at approximately the same level of the construct are not at equal intervals, suggesting that change in repertoires of measurement conceptions is not a matter of simple accretion over time. This is not surprising, considering the entanglement of professional development activity, tools, and instructional materials necessary to sustain the learning progression.

Formative Assessment and In-Situ Evidence of Student Learning

During instruction, teachers observed and recorded instances of student thinking and action consistent with particular levels and sublevels of a construct. Students also responded to formative assessments, usually administered in a paper-and-pencil format, during the course of instruction. Teachers used scoring guides to relate student responses to particular sublevels of a construct. Teacher judgments of student responses were entered into TOTs, and teachers reviewed selected items and student responses with their class during subsequent formative assessment conversations. Further evidence of student learning during these conversations was also entered into TOTs, so that formative assessments blended student responses to items and subsequent elaborations of learning that emerged during formative assessment conversations (Lehrer, 2021; Lehrer & Wilson, 2021).

Guiding instruction by monitoring conceptual development

Here we illustrate how a pair of first-grade teachers employed *in situ* and formative assessment to inform next instructional steps for length measure in light of children's conceptual development. In the fourth year of the project, teachers introduced children to foundations of length measure through direct comparisons of representations of common attributes of "height" and girth of a collection of squashes (pumpkins). The network of concepts related to comparing attributes directly is described in the first level of the length construct. Figure 8.5

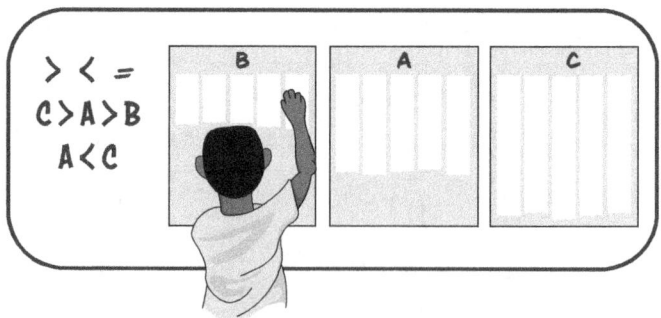

FIGURE 8.5 Student directly compares representations to order magnitudes of length

depicts *in situ* evidence of a student (shown via a drawing to preserve anonymity) directly comparing his representation of the height of one pumpkin (B) to that of another (A) to infer and subsequently to symbolize an ordered relation of magnitudes as A > B. Relations among magnitudes of height were subsequently extended to include another pumpkin, C. This performance is indicative of the first length construct level.

On the basis of this observation and others like it, as well as observations of student responses to a formative assessment, teachers were confident that students were in a position to embark on indirect measure of length by considering qualities of units that afford reliable comparison of magnitudes of different lengths. These qualities are the primary conceptual focus of the second level of the construct map. *In situ* observations such as "because if you have gaps, you are not measuring part of the line" were again buttressed by students' responses to formative assessments. For instance, one of the items on a formative assessment for Length unit 2 asks students in the class to signal with their hands (e.g., thumb up indicates agreement) which of three claims they think is right: "A person's height is 5 feet. Someone measured his height with foot-long units like this one (show a foot-ruler) but left gaps between the units (demonstrate with two rulers). Get ready to put up your thumb to show which idea you agree with: Put up your thumb if you agree that the measure made with gaps will be less than 5 feet. Who agrees that the measure made with gaps will be more than 5 feet? Who thinks it will be exactly 5 feet?"

"Someone else measured the height with foot-long units like this one but overlapped the units (demonstrate with two rulers). Who agrees that the measure made with overlaps will be less than 5 feet? More than 5 feet? Exactly 5 feet?"

In-class polling, confirmed by a video record of children "voting" on these ideas in one of the first-grade classes during the fourth year of the project, revealed that 80% of the children correctly predicted that when there are gaps between units, the measure will be less than 5 feet, but when the units overlap, it will be greater than 5 feet. In the conversation that followed, several children explained that if there are gaps between the units of measure, some of the length is not measured ("left out"), and if the units overlap, some of the space is measured more than once. Teachers accepted a correct prediction of the effects on measure as evidence that children understood these ideas. Thus, students who answered both questions correctly were scored at the second level of the length construct.

Satisfied that most children were making substantive progress in understanding properties of units that afforded consistent measure, the first-grade teachers shifted the focus of instruction to unit iteration and symbolization of unit intervals on scales of measure, performances indicative of the third level.

Representative *in situ* observations included a video record of a pair of students who predicted that the measure of a path in dowel rods would be less than its measure in clipboards (level 2, compensation relation) and then iterated a rod and clipboard to test their prediction (level 3). Other *in situ* observations included video of a disagreement between a pair of students about the location and identity of numerals as they constructed a 4 "pink-foot" tape measure. This disagreement was eventually resolved by writing the appropriate numerals at the endpoint of the unit interval (level 3). Student responses to a formative assessment, especially items probing multiplicative comparison of length (e.g., "times as long," level 3), further fleshed out children's grasp of unit iteration and its entailments.

Subsequently, teachers pivoted instruction to construction of half-unit and measure in half-unit to create measured quantities, conceptions associated with the fourth level of the length construct. In previous years, teachers regarded understanding of equipartitioning of a unit to create $\frac{1}{2}$ u and to construct measures in $\frac{1}{2}$ u through iteration as being at the edge of children's grasp, with comparatively few first-grade students exhibiting consistent understanding. However, with the resumption of daily in-person instruction during the fourth year of the project, first-grade teachers decided to address the construction of $\frac{1}{2}$ u and coordinated measure in $\frac{1}{2}$ u and u (e.g., $\frac{5}{2}$ u = $2\frac{1}{2}$ u) more comprehensively. Entertaining this possibility, teachers wondered: could a majority of first-grade students conceive of half-unit measure? Individual students' conceptions of half-unit iteration were inferred again from *in situ* observation complemented by children's responses to formative assessments (e.g., given a paper unit length, draw a line $\frac{3}{2}$ u long). Figures 8.6 and 8.7 depict *in situ* and formative assessment observations, respectively, in which teachers inferred that students grasped how to coordinate measure in half and whole units (level 4). (Both of these figures are facsimiles, drawn by an artist, of photos taken by teachers.)

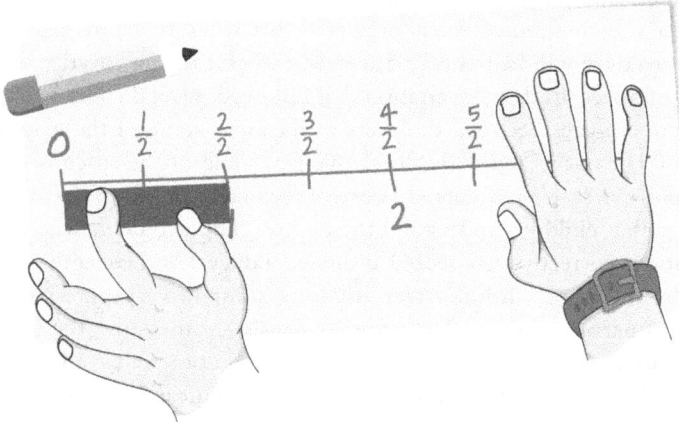

FIGURE 8.6 *In situ* observation, coordinating half- and whole-unit measure

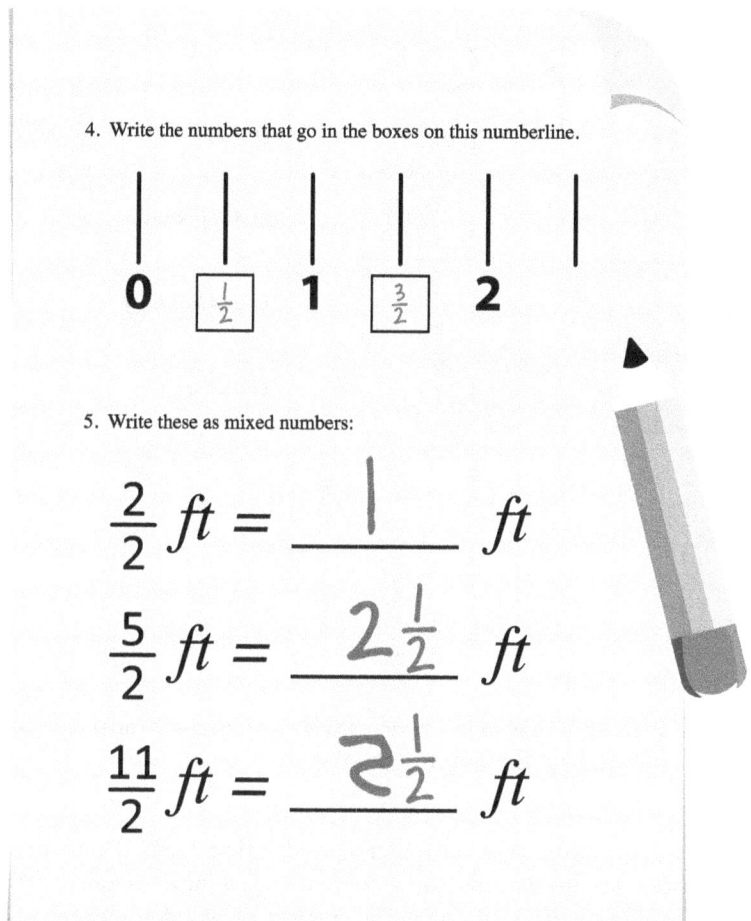

4. Write the numbers that go in the boxes on this numberline.

$$0 \quad \boxed{\tfrac{1}{2}} \quad 1 \quad \boxed{\tfrac{3}{2}} \quad 2$$

5. Write these as mixed numbers:

$$\frac{2}{2} ft = \underline{\quad 1 \quad} ft$$

$$\frac{5}{2} ft = \underline{\quad 2\tfrac{1}{2} \quad} ft$$

$$\frac{11}{2} ft = \underline{\quad 2\tfrac{1}{2} \quad} ft$$

FIGURE 8.7 Formative assessment observation, coordinating half- and whole-unit measure

To portray the time course of conceptual development, Figure 8.8 illustrates construct-centered growth in conceptions of length measure during the course of the school year in these two first-grade classrooms. Each region in the figure denotes the percentage of individual students observed at that level of the construct at least once.

Formative assessment and dialogic space

Formative assessment conversations open and sustain dialogues in which students elaborate their understandings of the mathematics of measure. Here we describe a formative assessment conversation that exemplifies how opportunities to learn can emerge during these episodes.

Construct Level:

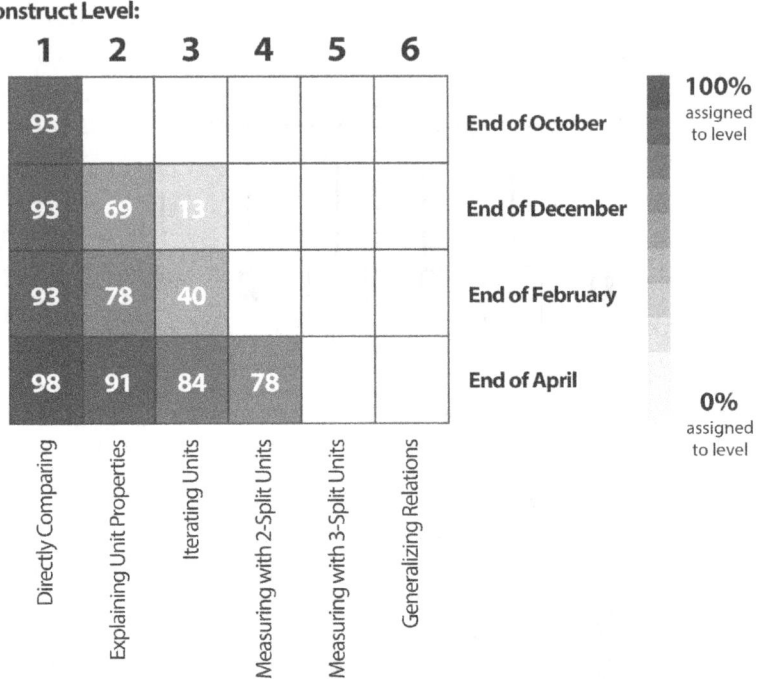

FIGURE 8.8 Percent of first-grade students who understood benchmark concepts of length measure over a school year

The assessment conversation under consideration focused on instruction in which students investigated how to find the volume of open prisms of varying dimensions, using the material support of wooden cubes with edge length of one inch. Many students filled the prism with cubes and counted all of them or skip-counted by composites (e.g., columns of cubes) to obtain a measure. This form of thinking is consistent with the third level of the volume measure construct. A majority of student solutions consisted of partial structuring of the prisms with cubes (e.g., by configuring cubes to align with two faces joined by an edge). This structuring helped them visualize the prisms as either horizontal or vertical layers of cubes. Volume measure was obtained by imagining the iteration of a layer, a form of thinking consistent with the fourth level of the volume construct. However, a few students ($n = 3$, 12%) proposed an alternative way of thinking—a dynamic conception of volume measure, generated by moving the area of the base of the prism, measured in square wooden-cube edges, through its height. (These students were apparently relying on the memory of their previous experiences of sweeping lengths to generate area.) The teacher used gesture to reanimate the formation of a unit cube by first sweeping edge by edge to generate an area and then creating a cube by moving the square edge through the third edge. She encouraged students to try out this novel sweeping conception (initiated at the sixth level of the volume construct) as they investigated new problems, and many did so. At the end of the class, some students related dynamic sweeping to multiplication and proposed rewriting a volume formula for the prism, $l \times w \times h$, as Area × Height.

After this class, the teacher administered portions of a formative assessment that included an item that asked students to find the volume measure of a 6 in × 4 in × 10 in prism depicted in a diagram. Nearly all the students calculated the volume measure correctly, even though the prior instruction did not incorporate diagrams; it had been carried out with three-dimensional materials. During the formative assessment conversation that followed, the teacher chose to focus further on dynamic reasoning.

She began by asking students, "Think about how you solved this problem. How did you start? Can you draw it on paper?" A student from one of the working groups replied, "We started with the 6 and the 4 ... we multiplied it." To provoke a clearer explanation, the teacher drew a 6-inch line on the board and iterated it four times, asking, "Is this what you did?" The student objected, "We are not making a line that is 24 inches; we are making 24 square units."

The teacher continued: "What do I have to do with this 6 inches in order to create a space inside?" (She gestured to indicate a region of planar space above the 6-inch line segment.) The student replied, "You use the 4 inches (gesturing to show vertical placement) to sweep it up." The teacher then drew the 6 inches iterated vertically through 4 inches and asked, "So what did we create there?" A different student responded, "We created 24 square inches." The teacher followed up with students' drawings to annotate the unit dissection of a rectangle to produce a grid of 24 square inches. She asked, "So now what? What do we do next?"

STUDENT: Then you multiply the 24 square inches by 10.

TEACHER: Ten what?

STUDENT: Ten inches, so you get 240 inches cubed.

TEACHER: When she says multiply by 10, what is she really doing?

STUDENT: She's sweeping it up (gestures vertical movement).

TEACHER: So she is taking it out to another dimension, this way (gestures a motion from the plane of the whiteboard into the space in front of students).

RL: So what is the volume after the first inch? If you are sweeping 24 square inches and you are going to sweep it through 10 inches, how much volume after 1 inch?

STUDENT: It's becoming three-dimensional!

After determining that the 24 squares had grown into cubes, the class turned to identifying the volume "when it goes up 2 inches," "when it goes up 3 inches," and, finally, "when it goes up 10 inches." To help students focus more directly on continuous generation of volume, RL invited students to consider the volume generated when the 24-square-inch base was swept through $\frac{1}{2}$ inch. Students worked in small groups on this problem, and several groups seized the initiative to extend the investigation to successive halving (e.g., $\frac{1}{4}, \frac{1}{8}, \frac{1}{16}, \ldots, \frac{1}{2048}$). Here students were deploying a familiar 2-split structure, initially developed while splitting lengths (see Chapter 3), as a way of coming to see that volume could be conceived as generated in ever-thinner and, ultimately, arbitrarily thin slices.

While students were working through the implications of ever-finer dynamics of generation, a parallel challenge posed was to consider how the prism would have to change to double its volume. Several students proposed doubling all of the prism's lengths but were disappointed to find that doing so resulted in a volume of 1920 cubic inches. Sarah interrupted to report that she had found the solution: "I doubled one dimension." When the teacher asked her to explain why this solution worked, Sarah replied, "If you double one dimension, then it's the same" (Sarah's gesture mimed iterating the volume of the original prism). The teacher repeated Sarah's gesture and elaborated, "So, Sarah said that if you double one dimension, you change the volume to make it twice as much." The class went on to symbolize this action in an expression: $480 \text{ in}^3 = 2 \times 240 \text{ in}^3$. Someone pointed out that the originally proposed solution of doubling each dimension, resulting in a volume measure of 1920 in^3, was actually *eight* times (rather than twice) the original prism's volume of 240 in^3. This was symbolized as $1920 \text{ in}^3 = 8 \times 240 \text{ in}^3$, which, as another student noticed, meant that the initial volume was $\frac{1}{8}$ times the volume of the (imagined) prism for which each side length was doubled.

During a subsequent class, students discussed their responses to other formative assessment items, including those presenting the challenge of finding the volume of a prism with fractional units of length measure for one of its edges, and novelties such as determining the volume of a hexagonal prism and a cylinder. During the course of this conversation, a group of approximately one-third of the class confidently asserted that, given the area of a base, the measure of the volume of a cylinder and the measure of a hexagonal prism were each the product of the area of the base "stretched through" its height. Other students grappled with this extension of the dynamic approach beyond the original context of rectangular prisms. Some were only now extending their initial fill-and-count strategies to partial structuring of prisms. While these groups continued to work, the teacher further challenged the first group, who were pushing at the frontier of dynamic thinking. To do so, she provided a physical model of an equilateral triangular prism and challenged the students to find its volume. The students asked how to find the area of the triangular base, and the teacher provided grid paper and a ruler and responded that this was part of the challenge.

As they began to grapple with this problem, one of the students visualized two equilateral triangles arranged into a rhombus. With this visualization in mind, the student dissected the rhombus into a square by first imagining and gesturing a cut of one of the triangles into two right triangles and then rearranging the cut piece to constitute a square. This meant that the area of the equilateral triangle had to be half of the product of the base and height (the vertical cut) of the rhombus.

Other students wondered what would happen if the triangle was not equilateral and were not convinced that the area of other triangles would also be half the product of the base and height. With support from another participating teacher, students began an investigation of triangles inscribed within rectangles and used dissection (e.g., see Chapter 5) to establish the generalization of the initial insight to the area of any triangular base. With a measure of the area of the base in hand, students again generated the volume by thinking of the area of the base as being pulled dynamically through its height.

The arc of this discussion exemplifies the purposes and advantages of formative assessment conversations in general. First, formative assessment conversations display for the teacher the range and variability of student thinking about the topic at hand. Before the conversation begins, teachers have reviewed and digested written responses from all the students in their class, but the written work does not always clarify what motivates a particular response. The assessment conversation provides opportunities to clarify children's thinking by soliciting their justifications. This knowledge is critical for evaluating instruction that has been conducted and also for planning next instructional steps.

Second, the formative assessment conversation is at least as important for students as for teachers. It is a forum for children to articulate and justify their thinking as they make these ideas public for the class. These opportunities to elaborate in detail and to communicate so that others understand can help children stabilize (or question) their thinking. Hearing ideas that others are proposing can generate cognitive change, especially when students are held accountable for understanding, revoicing, and/or questioning the thoughts of peers. When the classroom culture values variable solutions and strategies, children are more likely to be exposed to ideas that they are not yet quite ready to generate on their own.

Finally, this example demonstrates how a formative assessment conversation can pull the class forward into next near opportunities for learning. Indeed, the boundary between assessment and instruction is extremely permeable. Recall that the originally posed problem was to find the volume of a prism illustrated in a diagram, but as the formative assessment conversation proceeded, several opportunities emerged for children to extend their learning even further. These opportunities varied somewhat, depending on students' current ways of thinking.

For instance, in the conversational exchanges that we described, perhaps two or three children benefited from revisiting the relations and differences between length, area, and volume. Progress made by these students was primarily in structuring volume. For example, instead of filling a rectangular prism with cubes and counting them all, some of these children progressed to being able to line the length, width, and height of the prism to find its dimensions and using that information to visualize the internal space of the prism as structured into composite units. They were then able to explore the implications of taking any surface of the prism as the base, so that "layers" of volume could be visualized as either horizontally or vertically stacked.

Most of the students in the class were no longer tied to thinking of volume of rectangular prisms as discrete layers; instead, they communicated ideas about volume with dynamic metaphors and gestures (involving sweeping and stretching) that imply continuity. These children entertained thought experiments about creating smaller and smaller quantities of volume by moving an area through infinitesimal length (height). Continuity was also extended to stretching one or more dimensions of a prism, as students investigated the entailments of doubling volume. Additional investigations were devoted to finding the volume of a prism with one or more dimensions measured in fractional units. For example, if the base of a prism is 12 in^2 and its height is $3\frac{1}{2}$ in, what exactly does one do with the $\frac{1}{2}$ in? (Some children initially proposed that the volume would be $36\frac{1}{2}$ in^2.)

Finally, about a third or so of the class generalized dynamic images of volume as movement of area through height to find the volume measure of cylinders and prisms with other bases. As these examples demonstrate, an assessment conversation is not just the evaluation of past learning; it can also entice students into further learning, and these instances of progress in learning in this class were documented in TOTs. Figure 8.9 illustrates the distribution of individuals' ways of thinking at the conclusion of the second formative assessment conversation according to *in situ* observation and records of written responses to items. Notice the growth in dynamic thinking from its initial

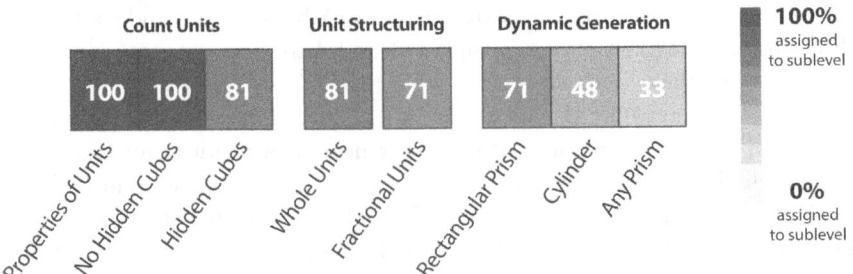

FIGURE 8.9 Percent of fifth-grade students who understood concepts of volume measure

seed to encompass a majority of the classroom in its initial forms (at level 6), if not yet in its generalization to cylinders and nonrectangular prisms.

Reflections and Prospects

The initial phase of research, conducted at Mallard Elementary School, supported researchers and teachers in developing a more comprehensive and detailed understanding of student learning about measure in the realms of length, angle, area, and volume. The findings from yearly flexible interviews guided the ongoing revision of curriculum and helped attune educators to the range of conceptions and strategies that youngsters employ as they reason about measure. To illustrate the flavor of the classroom studies conducted in ongoing waves within the participating classrooms, we also described a design study of students using a technology-based tool to investigate the generation of area via sweeping.

When the project shifted to its tryout phase at Sleeve Elementary School, we began to administer more traditional forms of summative assessment to individual children. These broad-scale assessments were assembled from an item bank that was developed and retuned during the initial phase of the project. The purpose of these assessments was to provide evidence of learning by comparing pre-instruction and post-instruction performance for participating students. In addition, during phase two we also experimented with new forms of assessment, conducted at the classroom level, that relied upon the professional judgment of teachers. Our colleagues at the University of California, Berkeley, are now investigating the feasibility of incorporating the student responses to summative assessments with the formative evidence of learning generated in classrooms. Our reasoning is that each form of assessment has different strengths, and a model that integrates both may be more informative than one based only on a single source of data. Such an integrated assessment provides grounds and practical means for accountability to the improvement of students' and teachers' opportunities to learn.

References

Brady, C., & Lehrer, R. (2021). Sweeping area across physical and virtual environments. *Digital Experiences in Mathematics Education, 7*, 66–98.

Horn, I. S., Kane, B. D., & Wilson, J. (2015). Making sense of student performance data: Data use logics and mathematics teachers' learning opportunities. *American Educational Research Journal, 52*(2), 208–242.

Lehrer, R. (2021, August). *Accountable assessment.* Invited address, Australian Council for Educational Research Conference, Sydney, Australia.

Lehrer, R., Kim, M.-J., Ayers, E., & Wilson, M. (2014). Toward establishing a learning progression to support the development of statistical reasoning. In A. P. Maloney, J. Confrey, & K. H. Nguyen (Eds.), *Learning over time: Learning trajectories in mathematics education* (pp. 31–59). Charlotte, NC: Information Age.

Lehrer, R., & Wilson, M. W. (2021, October). *Changing the face of accountable assessment.* Paper presented at the NCME Classroom Assessment Conference.

Papert, S. (1980). *Mindstorms: Children, computers, and powerful ideas.* New York, NY: Basic Books.

Pellegrino, J. W., Chudowsky, N., & Glaser, R. (2001). *Knowing what students know: The science and design of educational assessment.* Washington, DC: National Academy Press.

Sarama, J., Clements, D. H., Barrett, J. E., Cullen, C. J., Hudyma, A., & Vanegas, Y. (2021). Length measurement in the early years: Teaching and learning with learning trajectories. *Mathematical Thinking and Learning, 6*(2), 81–89.

Shepard, L. A. (2010). What the marketplace has brought us: Item-by-item teaching with little instructional insight. *Peabody Journal of Education, 85*(2), 246–257.

Wilson, M. (2005). *Constructing measures: An item response modeling approach.* Mahwah, NJ: Lawrence Erlbaum Associates.

Wilson, M. (2021). *Rethinking measurement for accountable assessment.* Invited address, Australian Council for Educational Research Conference, Sydney, Australia.

Wilson, M., & Lehrer, R. (2021). Improving learning: Using a learning progression to coordinate instruction and assessment. *Frontiers in education: The use of organized learning models in assessment,* Vol 6. doi: 10.3389/feduc.2021.654212

Wongkamalasai, M. (2018, September). *Positioning young children's visualizations of rotations as tools to define mathematical objects.* Paper presented at Spatial Cognition 2018, Tübingen, Germany.

9

TEACHER LEARNING

This chapter describes principles and formats of collaboration with teachers that were pursued to establish the learning progression in measurement. As Chapter 2 explains, the learning progression's focus on students' thinking required a partnership with teachers to establish (1) a shared conceptual frame for describing instructional goals and valued forms of teaching and learning—a collective professional vision articulated most succinctly in the constructs, but realized in everyday practices (Goodwin, 2018); (2) curriculum units designed to support teaching that is responsive to students' emerging grasp of core concepts and procedures in measure; (3) an innovative assessment system that aligns in-the-moment teacher observations of student learning with formative and summative assessments that collectively contribute to responsive teaching and to potentially more robust estimates of student learning; (4) a set of digital tools to help teachers detect, share, analyze, and interpret evidence of student learning; and (5) classroom and school-level professional development activity structures to support and sustain the progression. This fifth aspect of the project, professional development, played an integral role in the development of all of the other project components, because participating teachers were involved at every stage in informing ongoing revision of all program elements. The ensemble of these five strands of activity constituted an epistemic culture in which the learning progression was articulated (Knorr Cetina, 1999). By *epistemic culture* we refer to the enterprise of making, contesting, and revising knowledge, here with the aim of supporting, sustaining, and contesting knowledge about teaching and learning of measurement. This more expanded view of the work of a learning progression goes beyond the typical emphasis on evaluation of the empirical adequacy of the descriptions of typical patterns in conceptual development (though it does not neglect them) to encompass the

DOI: 10.4324/9781003287476-9

intertwining strands just noted (e.g., Alonzo & Elby, 2019; Confrey, Maloney, Nguyen, & Rupp, 2014; Lehrer, Kim, Ayers, & Wilson, 2014).

In this chapter, we explain two major aspects of professional development. First, we describe in further detail how professional development goals were embedded within and pursued hand in hand with program development goals. For this purpose, we focus primarily on participating teachers at Sleeve Elementary, although the work in early years at Mallard Elementary was also essential in informing development work on the components of the project. Our purpose is to illustrate how teacher participation with constructs, curriculum, and tools, in both everyday teaching and professional development formats, led to an iterative cycle: teachers informed the continuing redesign of the project components; at the same time, the constructs, curriculum, and tools became important agents of teachers' ongoing professional development, both in professional development settings and as teachers carried out their daily teaching responsibilities. Second, we follow this discussion of the interrelations between professional development and program design with evidence obtained from yearly interviews with teachers about the extent to which our goals were met.

Initiating and Sustaining a Teacher-Researcher Partnership

Table 2.1 in Chapter 2 describes the numbers of teachers at each grade level who participated in the partnership. At Sleeve Elementary, a core of teachers participated in all four years of our partnership there, although the nature of instruction and some aspects of our work together were altered by the global coronavirus pandemic. The pandemic foreshortened the 2019–2020 year of the project (the second year at Sleeve) by approximately two and a half months, and instruction during 2020–2021 included a mixture of in-person and digitally based formats. In-person meetings that year resumed only during the final month of the school year. During the fourth and final year of the study (2021–2022), two participating teachers were removed from classroom instruction and assigned to new roles in focused instructional support to mitigate the effects of the pandemic on student learning. A few additional teachers left the district or were unable to participate due to the increasing stress associated with managing pandemic conditions at school and at home. Consequently, during the final year there were 11 participants at grades 1–5, the mathematics coordinator, and, in addition, one of the teachers assigned to a new instructional role (this teacher does not appear in Table 2.1), for a total of 13 participants at Sleeve Elementary. During this final year, the district's mathematics coordinator also participated in most of the professional development. Her aim was to incorporate the material and professional practices of the learning progression into the district's elementary mathematics education.

Supporting Development of a Shared Professional Vision

As explained in Chapter 2, we collaborated with teachers to constitute forms of professional development in which common visions of teaching and learning could emerge—a shared professional vision (Goodwin, 2018). Professional vision played several important roles in sustaining the learning progression. Especially important was the development of a common lexicon (Mesiti, Artigue, Hollingsworth, Cao, & Clarke, 2021) and perspective about students' ways of thinking about core concepts of measure, grounded in shared understandings of how these ways of thinking were manifested in student talk, products, and activity. This framing of student learning facilitated a transition in the locus of instructional improvement from an individual teacher's effort in a self-contained classroom to a school-wide enterprise in which improvement could be amplified by collective endeavor and critique.

To support the emergence of a common lexicon and perspective about teaching and learning of the mathematics of measure, we introduced new conceptual and material tools and new formats of professional activity during which these conceptual and material tools could become integrated into professional practice. The premise was that if ongoing assessment of student thinking were woven into the fabric of instruction, along with an image of typical horizons of change (Ball, 1993), then teacher judgments of students' ways of thinking could lead to ongoing improvement of instructional practice and student learning (Copur-Gencturk, Plowman, & Bai, 2019; Gibbons & Cobb, 2017). Achieving this goal meant that teachers had to learn to register selected forms of student thinking as those forms emerged during the course of classroom activity. Moreover, based on these judgments, teachers then had to develop instructional practices that would leverage their knowledge of student thinking to improve the quality of instruction, so that assessment could contribute to instruction that was sensitive to how students thought about core concepts (big ideas) in measure. In the next section, we describe the conceptual and material tools that we designed to support the development of this unique form of professional vision. We then describe how these tools were deployed in new participation structures to support a professional learning community.

Constructs

As Chapter 2 explains, trajectories of conceptual development were encapsulated as constructs (Wilson, 2005) that describe the conceptual milestones, as articulated in previous chapters, in measure of length, area, angle, and volume. Constructs are structured so that the forms of thinking they describe become tangible and visible in classroom talk and activity and thereby can be recognized and acted on by teachers. Each construct is expressed in two forms: both as a *narrative* of conceptual development that is realized across multiple years of schooling and as a *table* that summarizes a progression through a series of discrete levels, each a major milestone of development. Narratives help teachers get a grip on the construct's story of development—its framing of change and continuity of core concepts of measure in each realm of measure. The narrative organization also poses milestones of conceptual development as a network of relations among ideas that collectively constitute the system of thinking labeled by the milestone. In contrast, the tabular format expresses each conceptual milestone as a level (e.g., conceiving of how properties of units affect a measure of length) constituted by a set of discrete concepts or sublevels, each of which can be viewed as contributing to the "big idea" of the level. Each sublevel is exemplified by observable instances of student talk and activity that can be registered by teachers attuned to that sublevel. For example, Table 3.2 in Chapter 3 illustrates the discrete constituents of the second level of the length construct, which describes how students learn to recognize and explain how properties of unit affect the measure of a magnitude of length.

The performances described at each sublevel of a construct are not only classifiers of student thinking but also objects of professional inquiry. For example, during a lesson conducted collaboratively by a group of teachers, the lead teacher displayed a line segment tiled by five paper clips. After the children agreed that the length of the line measured five large paper clips, the teacher next asked where on the line zero should be located, because she was aware that it is important to seek evidence that students perceive the need to symbolize the origin of measure as zero (this knowledge is classified as sublevel 3B of the length construct, *Symbolize/write starting point of measure as zero*). During the ensuing 20 minutes of conversation, students revealed several distinct ways of thinking about zero. Some argued that zero indicates the absence of extent, so the best location of zero was in the negative space at either endpoint of the line segment. Other students contended that the best way to think of zero was to incorporate it into the forward number word sequence that accompanied accumulating units of measure, so that the first of the five paper clips tiling the segment would be designated as zero, with a resulting measure of four, rather than five, paper clips. Another idea advanced was that zero is not really a number, so perhaps it should be reserved for segments that have extent but, as yet, no units of measure. The idea that zero could indicate the starting point of measure was also advocated by several students. The teachers treated each of these student perspectives as sensible (see Rotman, 1987, on the disciplinary challenges of zero), and as a result, the simple declarative description of this sublevel was elaborated for the observing teachers into a space of possibilities for zero and its symbolization, in which one alternative, the origin of measure, was highlighted by the construct. As the teacher who initiated the question remarked at the conclusion of the conversation, "This is mind-boggling." Recurrent teacher inquiry meant that a sublevel of a construct increasingly generated a penumbra of possible ways that students could think of an idea like the starting point of a measure, and over time, levels and sublevels in the constructs came to serve as signposts that flagged concepts often associated with an unexpectedly wide range of student thinking. As we will shortly elaborate, the development of these conceptual anchors among teachers was facilitated when members worked together to plan, share, and interpret instructional responses to diverse student thinking. In this way, over time, constructs were progressively interpreted by the teaching community not as lists of correct ideas, but as a lexicon of learning (Mesiti et al., 2021).

As we collaborated with teachers to investigate how students at Sleeve Elementary thought about ideas in measure, constructs were repeatedly revised, most commonly by clarifying aspects of an example of a learning performance and/or by editing its description. Because substantive revision of constructs involved trade-offs between the complexity of the construct and its use in real time to notice and act upon student thinking, more consequential changes were comparatively rare and reflected consensus by teachers about a need for change. One of the more ambitious changes was the development of a rational number construct to describe transitions in students' thinking about fractional-unit quantities and about fractions as operators on measured quantities. Descriptions of student thinking about fractions were originally embedded within the other constructs, but as teachers increasingly recognized the utility of measurement for student sense-making about fractions, we mutually decided that these forms of thinking should be lifted out and made more visible in their own right as constructs. Item response modeling, as described in Chapter 8, also contributed to revisions of constructs, but only in concert with the contributions of the teacher professional learning community.

Curriculum

As mentioned in earlier chapters, we designed instructional units, each of which included a sequence of lessons, in each realm of measurement to promote students' conceptual development along the pathway described by constructs and sublevels of constructs. The lessons included brief preambles of mathematical foundations and meanings of particular constructs to provide a general orientation for teachers to the disciplinary focus of each lesson. Notes to teacher users, which were embedded throughout the lesson, clarified the rationale for tasks and suggested how teachers might fruitfully orchestrate classroom dialogue in light of anticipated patterns of student thinking, so that the curricular materials were "educative" (Davis & Krajcik, 2005). Tasks and tools that students were intended to complete were designed to provide opportunities for teachers to observe student thinking in action.

Every lesson includes formative assessment items (Black & Wiliam, 1998, 2009) aligned with particular levels on the construct map. Moreover, every formative assessment item is accompanied by a formative assessment conversation guide that alerts teachers to ways of productively harvesting expected variability in student responses in ways that advance learning during subsequent classroom conversation about the responses that students generate to the items. Each of these conversation guides was developed during collective work with teachers. For example, one of the conversation guides addressed students' typical responses to Figure 8.1b in Chapter 8. Some students insist that the line segment labeled A in Figure 8.1b has area, and indeed, as a literal object, it clearly occupies a two-dimensional region of space. As the following paragraph from this item's formative conversation guide illustrates, the guide highlights this student choice as an opportunity to help students consider the distinction between a line segment as a literal object and as a representation of a length:

> If some students chose the line segment as having area, this is an opportunity to consider a line segment as a representation vs. a line segment as a literal object. As a literal object, a line segment has area, as can be seen by magnifying it with a document camera. It will look like a rectangle. But as a representation, a line segment simply indicates distance traveled, or length, so that the same distance can be represented by line segments of varying thickness. Students often find the equivalence of varying width segments very convincing for helping them grasp the segment's role as a representation, not as a literal object.

As teachers employed lessons to support learning, they occasionally authored supplements to the activities described in the lesson, and these teacher-generated supplements were included in a "teacher's corner," available to other teachers, so that the curriculum continued to evolve as we learned more about means to support student learning. As in other instances of the design of educative curricula, lessons were continually revised in light of ongoing evidence about learning and instruction (Davis et al., 2014). These revisions were made accessible to teachers digitally, so that curricular resources were augmented and refined throughout the years of the project.

Digital observation tools

As described in Chapter 2, teachers used a digital platform that we call TOTs (Teacher Observation Tools) to record evidence of student activity and talk and to associate this evidence to a particular sublevel of a construct. Corey Brady designed and refined TOTs in response to feedback from teachers concerning how their practices of assessment could be productively supported. To illustrate, we briefly describe an observation record made by a teacher about two students who were engaged in unit iteration during a lesson on length measure. The observation record included the following elements, all arranged on the same screen: (1) a photo showing how a pair of students constructed a length of 4 teacher-foot units, juxtaposed with (2) a teacher annotation of the photo (the teacher drew circles on the photo to indicate issues that required particular attention—in this case, pointing out the way students marked the endpoint of each unit as they iterated), and (3) a note about some of the limitations of the students' approach that the teacher planned to raise during subsequent classroom conversation. The record also shows (4) the sublevel of the length construct that the teacher felt best represented the children's activity in this example, along with assignment of the pair to this sublevel.

As teachers repeatedly categorized the ongoing flow of student talk and performance according to the levels and sublevels of the constructs, we observed steady growth both in teachers' understanding of the development of student learning and in their understanding of the constructs. As mentioned in Chapter 2, teachers could use TOTs display tools to visualize progress in learning at classroom, grade, and school-wide levels. As Chapter 8 explains, this use of TOTs provided a classroom-based source of evidence about learning that was incorporated into new assessment models and that contributed to teacher conversations about progress in student understandings.

Teachers also used observational records as instructional aids. For instance, observations were often displayed by the teacher to students to exemplify strategies that were discussed during follow-up class conversations. For instance, in one measurement lesson, students were working toward cutting a paper strip just as long as four iterations of their teacher's foot. Students initially expected that the lengths cut by different pairs of students would be identical because they were all of the same measure ("within a smidgen," as one student put it). This prediction reflects the value that measures should be reproducible by others in the community (Ford, 2010). Although many of the strips were, indeed, within a smidgen, some were noticeably longer, while others were noticeably shorter. How could this be explained?

At this point, the teacher presented and replayed her video and photo observation records of some of the iteration strategies that pairs of students had just used to generate a 4-teacher-foot length. After showing each episode of a strategy on the screen, the teacher reenacted the strategy just reviewed with a unit length so that students could more closely observe and evaluate the methods used by their classmates. Students noticed how some pairs marked off unit lengths by laying a pencil at the end of each iterated unit. Although this strategy was helpful for keeping track of the number of iterations, it had the unforeseen effect of introducing gaps between translations of the unit length, resulting in strips that were longer than 4 teacher-feet. Students recalled that leaving gaps between iterations resulted in unmeasured space. Another student remarked that the width of the pencil inserted at the end of each unit to keep track of the number of iterations wasn't very long. So why, he wondered, was the strip constructed with this iterative strategy so much longer than the others? The reason was apparent after the class tried out the strategy—that is, measuring while juxtaposing pencil widths between units. Students noticed that as the pencil widths accumulated, they produced a noticeable additional, unmeasured length. Another student strategy displayed by TOTs and discussed by the class featured inadvertent overlaps between the translated units. These overlaps resulted in lengths with measures less than 4 teacher-feet because some of the length was measured more than once. Although these conceptions of unit tiling had been encountered previously by the class, those encounters occurred in contexts where students had sufficient units to cover the entire length being measured. Now these ideas were recontextualized as students sought to generate reliable procedures for iterating a single unit to measure. Students' earlier understandings of properties of unit—including tiling and avoiding gaps or overlaps among units—served as resources for understanding how these properties were manifested in a new performance (that is, iteration of the unit).

The TOTs recordings were also important for the learning of teachers. As teachers repeatedly encountered evidence of variability in student thinking, they began to reconceive of learning not as a one-and-done achievement, but rather as the ongoing elaboration of previous understandings and the extension of that knowledge into new contexts over the course of development. Thus, constructs are not intended as stages or descriptions of stable performances. Instead of being thought of as the straightforward accretion of new ideas, change in children's thinking is often better described as a shifting distribution of concepts and procedures (Samara & Clements, 2009; Wilson & Lehrer, 2021). In sum, TOTs was used both as a record to inform teachers about students' conceptual development and as a public classroom tool that made student thinking available for review and critique by the entire classroom community. In this manner, TOTs was used both to describe and to support changes in student learning.

Activity Structures That Forge a Professional Learning Community

Shared learning constructs and tools were fundamental in helping to establish communal practices of instruction and assessment, both of which were realized in contexts of shared activity structures. Our emphasis on community was informed by its critical role in the development of the professional discourse necessary to support the improvement of instruction (e.g., Ball & Cohen, 1999; Desimone, 2009; Gibbons & Cobb, 2017) and by its role in generating productive norms for assessment (Horn, Kane, & Wilson, 2015). We engaged in several forms of activity to help teachers to notice and describe student thinking in the lexicon of constructs, to learn how to leverage the resulting knowledge of student thinking to support instruction, and to construct records of student thinking that could inscribe progress in students' ways of understanding measure. To meet these aims, we engaged in three primary forms of professional activity, all of which made abundant use of the conceptual and material tools just described.

Learning labs

We adapted math labs or learning labs to generate opportunities for teachers to learn together from and with students (Kazemi et al., 2018). During a learning lab, participating teachers collaborated to plan, instruct, and reflect upon student learning *in situ*—that is, as classroom instruction proceeded. The school instructional facilitator and a researcher assisted at every lab, and two or three labs were conducted each day for two days every month during the course of the school year. When first introduced to this format of activity, teachers preferred to work in grade bands of K–2 and 3–5, but they increasingly came to value the opportunity to observe a wider spectrum of learning. Eventually, teachers constituted K–5 teams to pursue longer trajectories of conceptual development. To illustrate our adaptation of learning labs to support the learning progression approach, we next describe a learning lab conducted with six teachers representing grades 1–4, the instructional facilitator, and a researcher. The phases of the lab included planning, instruction, and reflection.

During the *planning phase*, a second-grade teacher, JSP, proposed to conduct a lesson on unit iteration by challenging pairs of students to use a single copy of a standard unit, an SP (the length of her foot), to construct a tape measure that would be 4 SP long. Students were invited to use the following materials: a roll of white paper machine tape, an SP paper strip length, a pencil, and scissors. JSP situated the task as a follow-up to a sequence of activities during which children had considered properties of unit, especially unit tiling, and the roles these properties played in constructing accurate measures of length.

As the planning phase continued, JSP reminded the other teachers of several previous episodes of instruction so that they would be prepared to contextualize what they observed during the lesson. Previous instruction included investigation of the effects of gaps and overlaps in units on measure of three lengths, all of the same magnitude, as well as students' encounters with the challenges of unit iteration as they tried to use paper-clip units to measure a length. JSP also explained her goals—namely, to introduce children to symbolizing units on the tape measure scale. She shared her conjecture that continuous motion would help students make sense of the location of numerals at endpoints of unit intervals. Several of the other teachers endorsed the importance of motion, especially for helping students make sense of symbolizing the starting point of measure with the numeral 0. (As noted previously, students' concepts of 0 had been the subject of an earlier teacher inquiry.) The team then settled on the strategy of observing pairs of students as they constructed the tape measure and, as they observed, identifying and recording relevant forms of student thinking in relation to sublevels of the length construct. Teachers established consensus about the sublevels they were most likely to observe as pairs of students worked. These were primarily at the third level of the construct, and teachers set up the observation application, TOTs, so that they would be ready to associate instances of student thinking and learning with sublevels of this third level. The participating researcher proposed further that if there were time at the end of lesson, the group could consider extending a theme from a previous learning lab with older, fourth-grade students by engaging children in distinguishing multiplicative and additive comparisons, here comparing magnitudes of 4 SP and 1 SP. Participants responded that this topic might be productive but needed to be contingent on evidence of student progress in unit iteration during the lesson.

The *classroom instruction phase* provides opportunities to observe variation in student thinking and to consider the effectiveness of forms of instructional support. Observations of student thinking and thoughts about instructional support are recorded with TOTs assignments of student thinking to constructs, often elaborated with teacher notes and annotations of photos and videos. Teachers work both individually and collectively during instruction. Although there is a lead teacher or pair of lead teachers, any member of the group may interrupt ongoing activity by proposing "Pause," whereupon instruction pauses and the group consults together about some aspect of the ongoing activity. Teachers also may initiate co-teaching, usually at the invitation of the lead teacher(s). Pauses and co-teaching increase opportunities for collective inquiry and interpretation (Ghousseini, Kavanagh, Dutro, & Kazemi, 2021). As the learning lab begins, teachers spread through the classroom to maximize collective observation of the range of students' ways of thinking. This arrangement reflects previously established norms to benefit the lead teacher or teachers by assisting instruction and by documenting student learning more thoroughly than is typically possible within a classroom.

JSP led off instruction by invoking and reviewing episodes from previous lessons that children recalled. For example, she asked children to recount how and why lengths with measures of 9 C (a child's foot) and 9 SP were not the same and how this difference underscored the importance of using units where everyone had "the same exact understanding of a unit, so that when we say 9 feet, we know exactly what we mean by feet." Children also recalled strategies they had developed to keep track of endpoints of unit intervals to enable iteration (e.g., using fingertips, pencil points).

JSP then described the day's task: designing a 4-SP tape measure. As pairs of students worked together, teachers observed and recorded aspects of student thinking that were aligned with the length construct. Teachers also tried out different forms of support, most frequently asking students to "travel" unit intervals as a guide to locating numerals on the tape measure (e.g., 0 as a starting point, 1 at the endpoint of the first unit interval). A few teachers also asked students to compare the 4-SP length to a unit length, both additively (how much longer?) and multiplicatively (how many times as long?). As is normative, observers shared with JSP what they were seeing as she prepared to conduct a whole-group discussion after children's tapes were completed.

JSP initiated the whole-group discussion by proposing to label the first unit by placing a 1 at the midpoint of the first unit interval. She illustrated what she meant by holding up a single unit length and animating travel of "one whole unit" with her hand moving along some of the paper-strip tape provided to students. Several students objected that labeling 1 at the midpoint of the unit was misleading because "it's not the whole unit." JSP then invoked motion: If "grandma's house" is 1 SP from the start (0), "and I stop here (gestures to midpoint of unit interval) because that is where you labeled it, am I at grandma's house?" The class chorused no, and there was further discussion of continuous motion by analogy to a moving car (the class had used toy cars to model motion during previous lessons), which a student described as "traveling all the way, and not jumping like a frog." This student was contrasting discrete unit translation with continuous transition through a path length, and many other students indicated agreement.

JSP seemed satisfied that the students appeared to have a good grasp on iterating units and a rationale for symbolizing distances traveled with numerals marking the endpoints of unit intervals. Next, she gestured along the edge of the tape and asked, "How long is this?" Students responded that the measure was 4 SP. Here, JSP hesitated and signaled a pause, saying that she was not sure about how to phrase the next question and looking at the other participating teachers for assistance. Another teacher, SD, joined in as an instructor. She mentioned that she had tried out the comparison that JSP had in mind with the pair of students she observed. Holding up a unit-strip length and the 4-SP tape, SD asked, "Which one is longer? And how much longer?" One student responded "Three," and another explained, "If you take one, then there is three more to get to four." SD revoiced this difference, describing it as addition: "One of these (gestures to SP unit) plus three more equals -." [Here SD deliberately omitted the answer to the equation, because her immediate goal was to refocus students on the idea that the same situation can be modeled in an alternative way, that is, as a multiplicative relationship.] She then proceeded to alert children that "we're going to introduce some new language, *times as long*." She requested a choral repetition of *times as long*. Holding up the 4-SP tape measure, she asked, "The 4 SP is how many times as long as the 1 SP?" Student responses included one, three, and four. SD asked Brayla why she thought it was four, and Brayla replied, "'Cause if you use the 1 four times, it will be four," spanning her hands to indicate 4 SP. The lead teacher, JSP, asked Brayla if she could say more and prompted the class to recall "the time we talked about copies" as JSP held up an SP unit length. Students recalled a previous conversation that had occurred during their introduction to unit iteration, about copying a paper-clip unit length. JSP then copied an SP unit and glued the copy, with tape, to the previous copy, repeating the process to create a 4-SP length. Another student, Elizabeth, ratified that four copies of the SP unit produced 4 SP. SD followed up by inviting one of the students she had observed previously to come to the front of the room and make a length that was three times as long as 1 SP. With some contributions from other students, he went on to produce lengths that were three, two, and one times as long as 1 SP.

Then a third teacher, JH, said that she wanted to "put this into third-grade language, an equation." She wrote "1 SP," which JSP described for the class as the start of a number sentence expressing that 1 SP is 1 times as long as 1 SP. A student volunteered that the way to write this number sentence was 1 SP × 1 = 1 SP. JH followed up by asking, "What about two copies?" She again initiated the expression by writing "1 SP." Several students chimed in that 1 SP × 2 = 2 SP. The class proceeded in this manner to write expressions for three times as long and four times as long, using the language of copying to represent iteration, so that *times as long* and number of iterations corresponded. Yet another teacher, MC, playfully suggested, "We did not make a ruler longer than 4 SP, but what if we had five copies?" Another teacher, KB, chimed in, "You mean just do it in their brains?" MC laughed and responded, "Yes, it's crazy." MC good-humoredly described this as fourth-grade thinking. There was now palpable excitement among students, and someone suggested that it must be 1 SP × 5 = 5 SP, "because you already know the other answers; you just have to add another number." MC responded with "Oh, my gosh!" and asked about 10 copies. A student exclaimed, "or 100!" Students quickly wrote corresponding expressions, and as the end of the period arrived, JSP told the researcher that there was time for one more question. He asked how to express five copies of 2 SP, extending the discussion of *times as long* to composite units. JSP spontaneously elaborated by holding up 2-SP units taped together and gestured iterating or copying the composite. Students quickly completed the expression 2 SP × 5 = 10 SP and then started extending it to 2 SP × 10 and the like, accompanied by a swell of conversation and animation about the possibilities of expression. The extension to symbolic expression initiated by JH was not anticipated during planning, but with mutual bootstrapping by the team, it turned out to be very productive.

The aims of the *debriefing and reflection* phase of the learning lab are contingent on what the group notices during instruction, but usually include sharing teacher observations about student thinking and relative effectiveness of forms of instructional support. Debriefing also includes agreement about next steps and/or modifications of the instruction that will be tried out by other teachers, especially those at the same grade level.

In this learning lab, teachers shared what they noticed about student thinking during their work with pairs of students, which they had recorded with TOTs. Most of these ideas also emerged in the class conversation. As teachers shared, they used the same language as had been employed in the construct descriptions, although without explicitly mentioning sublevel (e.g., 3B) labels.

The teachers' reflective conversation indicated that they had common understandings of the students' ways of thinking described by the construct.

As teachers continued to share observations of students in the learning lab, they asked more intently about students' previous history of learning about tiling units, and JSP shared some forms of activity that had proven useful to support student learning. The conversation highlighted the apparent accessibility to these young children of the multiplicative comparison of the magnitudes of unit and measured length and revisited the importance of consistent linguistic support, especially the phrase *times as long* and the contrasting additive comparison language *how much more*. Teachers and the researcher also considered the advantages of different formats of expression for multiplicative comparisons, with the researcher favoring writing expressions in the form of "number of copies × measured quantity" so that the measured quantity was always in the position of the quantity being operated on. The follow-up debriefing also revealed some of the pragmatic challenges students faced in cutting paper and creating space for inscribing numeric symbols on their tape measure. Designers of tape and ruler measures often struggle with this problem, with the result that 0 is often not coincident with the beginning of the instrument.

The conversation shifted from immediate next steps to composite and part-unit iteration, goals relevant to longer-term development. Teachers asked, in light of the effects of the pandemic on children's histories of learning, what was reasonable to expect from students in grades 1–3 for the rest of the year? This conversation continued for approximately half an hour and touched on which conceptions of fractional quantities and of fractional operators on measured quantities might be accessible to students across the grade span. The possibility of multiplicative comparison stimulated further teacher inquiry. First-grade teachers declared that they were not used to thinking this way but decided to lead an exploration with first-grade students to learn whether younger children could make sense of this type of comparison, an investigation taken up in a subsequent learning lab. Ways of supporting children in multiplicative comparisons of measured lengths were encapsulated in a "teacher's corner" contribution to the curriculum materials—a further illustration of the interconnected nature of professional learning across formats of activity and material (e.g., Ehrenfeld, 2022). The orientation in this learning lab toward possible trajectories of long-term development was an index of the growth of the group's collective professional vision, which increasingly focused on inquiry that spanned time and grade levels.

Mathematical investigations

Another form of community building is accomplished by involving participating teachers in *mathematical investigations*—that is, group inquiries about the mathematics of measure. For example, teachers explored properties of dynamic measures of space, such as how a length can be viewed as motion along a path, how an area can be generated by a length moved through a second length, how volume can be generated by an area moving through a length, and how an angle can be conceived as a directed rotation. They also considered how to help make fractions like $\frac{7}{3}$ more intelligible to students as measured quantities ($\frac{7}{3}$ u) of a distance traveled and how measurement could be employed to interpret arithmetic operations with fractions, especially multiplication and addition. These investigations were most often conducted in response to teacher requests during multi-day summer institutes but were also a component of the learning labs when the need arose, usually during the planning phase.

Communal critique

At the close of a day of learning labs, we jointly examined evidence of student learning being generated in classrooms, with an emphasis on helping others access the basis of evidence for assigning a student's understanding and performance to a particular level of the construct. This process was compared to auditing a tax return. We also used TOTs to consider progress in student learning at grade levels and across grade levels, so that we could visualize school-wide patterns of development. These visualizations instigated conversations about the aspects of instruction that needed further attention. During these conversations, teachers sometimes recommended changes to conceptual tools and the observational tool, TOTs. As agreement about evidence and how to use TOTs most effectively became increasingly shared, and as we shifted to conducting learning labs with teams that spanned the grades, these conversations became embedded in learning labs, with the school's instructional facilitator brokering among the groups involved in the learning labs.

Investigations of Change in Professional Practice

Researchers conducted yearly interviews to ascertain how teachers tended to interpret their participation in forums of professional development like learning labs and math investigations and to hear teachers' impressions of the conceptual tools and materials available to support teaching and learning. We also sought to ascertain the extent to which teachers were adopting a common frame for noticing and responding to student thinking about measure. To pursue these goals, we elicited what teachers noticed about videos of selected episodes of instruction and further asked how they cultivated student learning of "big ideas" in the course of their classroom instruction. We also investigated what teachers took as evidence of student learning in their TOTs records and the extent to which an independent observer of each observation record would agree with the teacher's classification of student performance with respect to the relevant construct.

Teacher Noticings During Video Episodes

At the outset of our collaboration with teachers at Sleeve Elementary, teachers viewed three episodes of classroom teaching in measurement, one at a time. We began by asking teachers to tell us what they noticed about concepts of measure and about the instructional practices that they viewed, as they were viewing each episode and/or immediately following their viewing (Sherin, Jacobs, & Philipp, 2010). We solicited teacher noticings again at the end of the first year of our collaboration. We conjectured that transitions in what teachers noticed about these episodes of classroom instruction would constitute evidence of the growth of professional vision, a common framing of key concepts in measure and instructional practices that could support student development. Our research questions were as follows:

1. Which concepts of measure did teachers notice as they viewed each episode? Our rationale was that concepts of measure are an essential lens for interpreting classroom events in which students are learning about measure.
2. How did teachers characterize instructional practices? Our rationale was that noticing *how* instructors are supporting learning bridges core mathematical concepts with specific forms of teacher support for these concepts.
3. Did teachers more readily notice core concepts and related instructional practices on the second occasion of review, when compared to the first? Transitions in noticing were taken as indicators of the growth of shared professional vision.

Video episodes

The video episodes were constructed from previous records of measurement lessons, one conducted at an elementary school in the Midwest and the other two at Mallard Elementary. In each of the three video episodes, a teacher orchestrated classroom conversation to promote students' understandings of measurement concepts and practices. These concepts and practices represented those identified by our research as essential and generative for students' long-term learning about measurement. Episodes faithfully reflected teacher-student and student-student interactions but were condensed somewhat to serve the purpose of eliciting video-stimulated noticing. One of the episodes was drawn from a grade 1–2 class, one from a grade 3 class, and the third from a grade 5 class, so that cumulatively the episodes reflected the range of grades taught by the participating teachers. Duration of the episodes ranged from approximately two to six minutes.

The first episode depicted first-grade students using unmarked paper strips to measure, compare, and order both the height and the girth of several pumpkins. The task was designed to engage children in foundations of measure, including the need to identify (e.g., what attributes can be measured?) and define (e.g., what is meant by height?) measurable attributes and to consider the role of a common origin of measure (e.g., where do we begin measuring from?). The episode began with brief glimpses of children's measurement activity and then transitioned to a discussion that occurred after pairs of children had completed their measures. The teacher in the video invited the class to compare the lengths of paper strips submitted by different groups as representations of the height of the same pumpkin. These strips were affixed to the wall but were deliberately arranged by the teacher so that their endpoints were not aligned. As a result, it was difficult to compare these representations of height, because the strips did not start from a common baseline. As the discussion proceeded, the first-grade students in the video came to appreciate that the strips could be effectively compared only if they were aligned along a common baseline. The teacher also highlighted the variability in the lengths of the strips generated by different small groups to indicate the same pumpkin's height and asked what might have contributed to this variability.

In the second episode, third-grade students investigated how to find the area and perimeter of a figure depicted as a C-shaped, irregular polygon with some values for the measures of the lengths of the sides missing. The task featured in the video is structured so that multiple student strategies for finding the measure of the area are possible. One can dissect the figure into three constituent rectangles, accounting appropriately for overlapping regions. An alternative strategy is to construct a segment that establishes a large rectangle by joining the endpoints of the upper and lower portions of the C shape. Then the area of the rectangular portion not bounded by the C can be subtracted. The task also provides opportunities for students to distinguish between area and perimeter, to consider units of measure, and to reason about properties of rectangles that could be used to determine the lengths of the unlabeled sides in the figure. The students shown in the video episode already had a prior history of viewing perimeter and area dynamically, as generated by motions in one and two dimensions, respectively. Accordingly, during the episode, the teacher in the video evoked a class history of motion in one dimension to animate travel along the perimeter path of the figure. She also employed a sweeping hand gesture to indicate portions of the area of the figure generated by moving one length through another. She reminded children about definitions of properties of a rectangle to provoke their thinking about necessary relations between known and unknown side lengths in the figure. As the conversation proceeded, she annotated the figure with dotted lines to highlight potential dissections into rectangles to promote consideration of the dissection strategies.

The third episode portrayed a fifth-grade class interpreting the meaning of a familiar formula for the measure of the volume of a prism in which length, width, and height were measured in units of inch (10 in, 5 in, and $3\frac{1}{2}$ in, respectively). Class conversation was directed toward justifying why the formula generated a measure of volume. The teacher in the video initiated the discussion by asking each student to make a drawing to show what length (10 in) × width (5 in) meant to him or her. Subsequently, she invoked students' previous experience with generating area by sweeping one length through another. She guided students in annotating a portion of the area generated by coordinating units of length measure to constitute square units (e.g., marking intervals of 1 inch on each side). The episode continued as the teacher displayed selected student drawings intended to illustrate length × width × 1 in. Building upon these student drawings, the teacher animated the volume dynamically as a sweep of the area of the base through the first inch of the prism's height. She further illustrated the product of this dynamic generation by presenting a 50-cubic-inch model that she had previously constructed. The episode concluded with conversation about student drawings intended to illustrate length × width × 2 in. The teacher again animated dynamic generation of volume and illustrated the product with a 100-in^3 prism composed of two 50-cubic-inch "layers." Throughout, the teacher anchored the symbolic expression of multiplication to dynamic generation of space and to tangible objects such as drawings and cubic arrays. The interview participants were asked to anticipate how the teacher might go on to support student learning about measures of volume that involve the fractional portion of the height.

Interview procedure

The first noticing interview was conducted at the initiation of our professional development collaboration, and the second was conducted at the end of the first academic year. Interviews were conducted one on one with a trained interviewer, who followed an interview script with branching probes. Each teacher viewed the three classroom video episodes in the same order, from episode 1 to episode 3. Participants were told in advance that they could pause the video at any point and then replay or continue to play the video, with the aim of describing what they noticed about the teacher's instruction and about students' thinking related to measurement.

After watching each episode, teachers were first asked what they noticed about "big ideas of measurement" ("What important concepts about measurement are students engaging with?") and then asked about the instructor's actions to support learning ("What, specifically, do you think the teacher is doing to support student learning?"). The interviewer sometimes interjected to ask a participant to clarify something she had said or to provide further detail. Each interview took about 45 minutes to complete.

Interview analysis

Responses to the interviews were video-recorded, transcribed, and audited to ensure fidelity of the transcript to teacher and interviewer turns of talk. Each transcript was first segmented into portions of talk in which a participant mentioned a concept in measure (e.g., "units, so that you don't forget your unit") and/or a teaching practice (e.g., "She's asking a lot of questions"). An initial coding scheme was generated by open coding (Strauss & Corbin, 1990) segments of the transcripts of two randomly selected Sleeve teachers and two more experienced teachers at Mallard Elementary who also participated in the interview. These initial coding schemes were also guided by our prior anticipations of the concepts of measure and of teaching practices that might be evident in the episodes. This process resulted in 497 codable segments. The codes were iteratively refined as they were initially applied, and a decision was made to allow multiple codes to apply to a single episode if they were appropriate. Axial coding resulted in three classes of codes. One class differentiated among concepts of measure mentioned by participating teachers, such as origin of measure. The second referred to general, domain-independent teaching practices such as "She asked a lot of questions" or "Calling on other students to help answer." The third referred to teaching practices that were linked explicitly to support of learning of measurement concepts, such as "and she had them do the hand motion, the perimeter, just to help them remember the difference between the two, because that's very tricky also." Coding was applied by two coders to segmented transcripts for the 11 participants who responded on both occasions of interview administration. A second random sample of 20% of the events was recoded, resulting in 86% exact agreement between coders. Most of the inter-coder disagreements involved segments in which one of the coders overlooked one of the multiple codes that could be applied to that event.

Interview results

Every instance of teacher noticing about instructional practices was counted across all three of the video episodes. At the outset of instruction (first interview), teachers most often noticed domain-general practices, which accounted for 54% of noticings about instructional practices. These included instructors' questions ("She's using questioning, and the questions I see were … those higher-level questioning techniques"), instructors' support for student agency ("encourage other students to build upon the thinking of another child"), and instructors' use of materials or tools to support student learning (e.g., "She's using a lot of visuals"). These instances were indicators of a positive discourse environment that was visible in each classroom teacher's interactions with students, but these noticings of instructional practices were not connected to explanations of how the instructional practice supported learning of a particular concept of measure. In contrast, in the second interview, domain-general noticings of instructional practice decreased to 13% of the total noticings of instructional practice, and the propensity to notice concept-specific instructional practices nearly doubled. In addition, teacher identification of measurement core concepts increased by 28%.

Table 9.1 illustrates, for each of the three episodes, what teachers tended to notice about core measurement concepts and concept-specific instructional practices. Entries in the table indicate the percentage of the 11 teachers who

TABLE 9.1 Transitions in Teachers' Noticings

Concept/practice noticed	At the onset	One year later
Episode 1 *Directly Comparing Heights and Girths of Pumpkins*		

First-grade students compare the lengths of paper strips generated by different groups to represent the height of the same pumpkin.

Concept/practice noticed	At the onset	One year later
Concept: Define attribute: "You have to establish what it is you want to measure."	91%	100%
Direct comparison: "Not (yet) units"	18%	82%
Origin of measure: "Like the same start point"	9%	64%
Practice: Highlight variability: "She asked why the measurements were not all the same."	45%	91%
Problematize comparison: "She was putting them at different heights so hard to compare."	9%	100%

Episode 2 *Finding Area and Perimeter of an Irregular Polygon*

Third-grade students consider how to find the area and perimeter of a C-shaped polygon figure.

Concept/practice noticed	At the onset	One year later
Concept: Unit: "Defining the unit that is used"	91%	73%
Rectangle properties: "Opposite sides equal"	64%	73%
Dynamic generation of length and/or area: "Times, that's where sweeping comes through."	36%	55%
Differentiation between area and perimeter: "Length is distance traveled, area is space inside."	36%	55%
Dissection of area: "Break it into rectangles."	18%	82%
Practice: Highlight defining properties of a rectangle: "What do we know about a rectangle?"	45%	55%
Appeal to dynamic motion: "She talks about when Ranton travels each step is the same."	27%	82%
Annotate figure: "She's drawn the lines to show them the whole figure."	9%	82%
Gestures to support learning: "She had them do the hand motion, the perimeter, to help them remember the difference" (area, perimeter).	36%	82%

Episode 3 *Interpreting the Meaning of a Formula for Volume Measure*

Fifth-grade students interpret the meaning of a familiar formula (length × width × height) for the measure of the volume of a prism.

Concept/practice noticed	At the onset	One year later
Practice: Appeal to dynamic motion: "Show me sweeping through one layer."	55%	91%
Tangible model supports visualization of unit, composite unit (layers): "She created the figure with the cubes and she actually showed them where those squared units were at, that seemed pretty powerful."	55%	100%
Elicit student drawings: "And then having them draw it and show, where's the 50? Where's the 2?"	18%	91%
Highlight unit: "Why is it 50 inches squared? Where did that come from?"	27%	55%
Link actions to symbolization $(l \times w \times h)$: "They are helping her write the equation that goes with it."	36%	100%

volunteered a particular core concept or a concept-specific form of instructional practice at least once during that episode. For instance, a teacher viewing the first episode might notice the core concept of origin of measure multiple times, but for that teacher, that idea would be counted only once. We set 55% as a threshold to indicate that a majority of teachers noticed a particular concept or related practice during either interview. For the third video episode we focused our analysis on noticings of instructional practices intended to help students make sense of the way the formula for finding volume works.

Table 9.1 shows substantial change in teacher noticings. These findings collectively suggest the emergence of a common vision within the participating teacher community of core concepts in measure and understanding of how instructional practices can promote learning about measurement concepts. For example, during the initial viewing of the first graders measuring pumpkins, only one teacher noticed how the instructor in the video was problematizing a common origin of measure to help children grasp its importance in comparing their measures. In contrast, after one year, all teachers noticed how the instructor made this core concept more visible to children by deliberately failing to align the strips at a common starting point. Teacher noticings also reflected increased sensitivity to the importance of dynamic metaphors of motion or gesture to support student learning of area and volume. Participating teachers were more likely at the second interview to mention the use by the instructor in the video of representations to support particular ideas. For example, a teacher viewing the second episode noticed how an instructor's annotation of the diagram helped students reconstitute the drawn figure as a more familiar figure, a rectangle: "Then she went into the kids realizing that it was like a rectangle, like it could be, like, a rectangle if you added that imaginary line there." (Reconstituting the figure as a rectangle enabled a dissection that involved subtraction of areas in the problem posed.)

At the end of the first year, coincident with the second wave of noticings, we included an additional task, posed after the discussion of all three video episodes. Our aim for this task was to probe whether and how teacher noticings about core concepts and instructional practices in the videos would be reflected in their interpretations of their own teaching. We were especially interested in whether teachers' accounts of their support for student learning would be portrayed as generic, domain-independent practices or whether they would instead relate specific instructional moves to goals for helping students understand particular measurement concepts. Therefore, after each teacher participating in the interview completed the discussion of the video episodes, she was next asked to describe two of the most important ideas about measurement that her instruction had focused on during the past year. After replying to this question, the teacher was further asked to explain how she had supported student learning of these concepts in her classroom. This task was designed to pursue the following questions:

1. To what extent would concepts that teachers noticed in the videos, such as properties of unit and dynamic images of length and area, be mentioned as core concepts promoted in their classrooms?
2. To what extent, and in what ways, would teacher accounts of instructional support for children's learning mention relations between specific teacher actions and aspects of the concepts mentioned as foci of instructional effort?

The interviewer began this portion of the interview by asking, "As you worked this year on measurement, what would you say were the two most important ideas that you helped children understand about measurement?" After the teacher's initial response, the interviewer posed a series of related questions: "Please tell me about a time when you helped children understand _____. What did you do to help them understand _____? What were some of the challenges that you experienced? Which choices that you made seemed to have the biggest impact on your students' learning?"

Teachers' responses were transcribed, and the transcriptions were compared to video records to assure their accuracy. Each transcript was then examined to determine how important ideas of measure were characterized by the participating teacher and how the she had helped children understand the important ideas that she had identified in the videos. Transcripts were coded axially to reflect emerging themes related to teachers' identification of important ideas. Thematic big ideas coded were about (1) properties of units, such as tiling and/or unit iteration; (2) dynamic generation of a length, area, volume, or angle; and (3) extension of measure to fractions or operations on fractions. Accounts of how the teacher supported student learning for each of the two big ideas nominated by that teacher were coded as either specific to measure or domain-independent. Episodes in which the teacher specified how a form of instructional support contributed to student learning about a big idea were classified as instances of specific support, identified by relating teaching actions to student learning about an idea in measure. Connectives were often causal (so, because, helps), temporal (and then, before, after, has been), or adversative (it's better than, on the other hand). For example, one teacher mentioned how highlighting a unit of length swept through another unit of length to generate a unit of area helped students come to see how dynamic generation of an area was related to its measure in units. In contrast, accounts coded as domain-independent were those in which the teacher recounted forms of activity that supported learning, such as questioning or modeling, but did not explain how that question or model promoted learning of a measurement concept she had identified as important. Some teachers also mentioned how their support for student learning represented a shift from previous practice. For example, one teacher said, "In the past, it's been more like 'This is how we do it' … but when they see for themselves what happens with gaps, with overlaps, with using different units, I think they understand it better than me making a list." These mentions of shifts from previous practice were also represented during open coding.

As with the video noticings, we report here the responses of the 11 teachers who had participated in both waves of the interview, in order to triangulate their responses to the videos with their remarks about their own classrooms. Each transcript was coded first for big ideas mentioned. Then, each teacher's retelling of instructional support was bracketed in the transcript as an episode and coded as either specific or domain-independent.

With respect to big ideas in measure, all but one teacher nominated dynamic generation of a length, area, volume, or angle as an idea worthy of instructional attention. Length as travel along a path was mentioned most often (64% of teachers), and a majority (55%) of teachers endorsed motion as a big idea that encompassed two or more strands (e.g., length and area). Many of the teachers (64%) also mentioned properties of unit as important, especially unit iteration. Extensions of measure to fractions were cited by 45% of the teachers.

Teachers recounted 46 distinct episodes of supporting classroom instruction, ranging from three to seven episodes each, with a median of four. Only a few episodes (15%) focused on domain-independent practices that supported learning. In one such episode, a teacher recalled helping students differentiate between surface area and volume by engaging in questioning and self-explanation: "I think just questioning all of that. Just really having them articulate what's going on in their head. I think that was the biggest thing. I had to allow them the time to work … also questioning their thinking and then asking them, 'Well, what would you do next?'" In contrast, most of the examples that teachers offered were focused on instructional moves that supported specific conceptual ideas in measurement. Many of these examples invoked motion, either literally or metaphorically, to help children think about a concept. For example, one teacher recalled how encouraging children to think of a unit as a distance traveled helped them think about a length measure as an accumulation of these unit distances: "One specific example I remember is using a little matchbox car. Actually, using that car to travel to that one inch, and using questions like 'Okay, is this one inch yet? Are we there?' All along the way until you get to that specific 1-inch measure, and then you can say, 'Now this entire thing is one inch. Okay, so how would we travel? Show me how to travel ten inches.'" Another teacher related length unit iteration as motion to help children think about fractions greater than one: "like the iterations of … of traveling [gestures iteration] has been helpful. It's always been in the past like when you get to five-fourths they're kind of like 'Ah, what do I do now?' And I feel like this year, it was not that way." Other teachers referred to bodily motion as helpful for developing understanding of angles-as-turns: "Well, we just had them stand up and use their body, so we had them do one full turn, then half a turn, and they started to understand what a fourth of a turn was or then what a third of a turn was when we did triangles. So just using their bodies [helped]." References to motion were also extended to area measure. A teacher recounted how moving one length through another helped students understand how units of area could be generated: "And the activity where they used the squeegee just really helped them see what they were covering [gestures a sweep], and when they swept it across to see those unit squares inside the shape. I've never really talked about the sweeping before, so I think that was really, really beneficial."

Many of these teacher accounts of instructional support mentioned the use of representations. For example, a teacher reported how using nets (2-D representations of 3-D structures) helped students relate surface area to volume, and another related how moving from physical to cybernetic manipulatives improved students' understanding of the dissection of area. Most of the teachers (82%) also indicated that the instructional practices they were recalling were shifts from their previous practice.

Teacher Perspectives on Professional Development

At the end of the second year of the project at Sleeve Elementary, which was truncated due to the pandemic, we asked all participating teachers ($n = 15$) about their impressions of the primary forms of professional activity in which they had participated. Unlike the previous two occasions of the interview (that is, at the onset of the project and the end of year 1), the conversation was digitally mediated via Zoom. Questions about forms of professional activity shared a common stem: "I'm going to mention some of the different ways that you have been working with other people on the project. And I'm going to ask you to tell me your impression of how well each of these ways has worked or has not worked for you." Then the interviewer asked about a particular form of activity. For example, for mathematical investigations, the interviewer said, "Sometimes with other teachers and with the research team, we explored mathematical ideas together, such as ways of thinking about measuring length or area or multiplying fractions. Can you think of a time when you found this kind of exploration of mathematical ideas productive and helpful for you?" A related question was "If we could do more exploration of mathematical ideas together, would that be beneficial for you?"

Teachers' responses to each question and its follow-up probes were transcribed, and transcriptions were compared to video records to assure their accuracy. Transcripts were open-coded, followed by axial coding to identify emergent themes. After these themes were identified, each transcript was recoded to identify instances of each theme in the transcript.

The sections that follow address what teachers had to say about their particippation in learning labs, mathematical investigations, and after-school, wholegroup conversations about the learning progression.

Teachers' views of learning labs

Teachers were asked, "If you were going to explain to a new teacher what usually happens in learning labs, how would you describe it?" All of the teachers indicated that learning labs were opportunities to *examine student thinking*. As one noted, "So, what usually happens is you get to see a lot of different thinking. So I really love that; it's one of my favorite things to do. To see how the kids are thinking. So, I would tell a new teacher to make sure they walk around, visit, ask questions. You'll gain so much more just by listening to the way the kids are talking about their math." Sixty percent of those interviewed further said that learning labs provide a window into the *development of student thinking*. As one teacher put it, "But it was actually way more beneficial for me when I could maybe see a first-grade lesson and then also like a third- or fourth-grade lesson, because I see where they're coming from and I see where they're going." The majority of teachers (53%) referred to learning labs as opportunities to formulate and try on different potential *images of practice*. For example: "I love going into other teachers' classrooms, because I feel you pick up at least multiple things. You go back and you're like, 'Okay, I want to get in my classroom, and I want to teach that lesson now, too, because I wasn't thinking of it that way.' So, just being able to see the different teaching styles and their questioning really helps."

Most teachers (85%) cited the collaborative, dialogic nature of the learning lab experience. For example, one mentioned, "It may be that you start off the lesson and then someone jumps in and asks a question or someone else says, 'Would you like me to notate what you're doing on the board up here?' So, it's more of a team effort as we all go in together." Another teacher stated, "You can look at it in different ways and learn from each other. There have been many times we've been together and someone said, 'Well, you know, I just thought about it like this,' and that's so much … And you have this, oh, that's so much more simple, so it's the same thing with teachers as it is in the classroom for kids." Eighty percent of the teachers indicated that pauses initiated during a lesson increased opportunities for dialogue among colleagues. As one teacher noted, "I mean, I think the biggest thing is just really getting someone else's idea. Or just the thought of how they're thinking about it. Because I might think about it one way, and another teacher might have thought something different. Or even if a student is explaining something and I'm not understanding what they're trying to say, another teacher's probably understanding it." Another teacher mentioned benefits to students as well: "I think, well, just outside of us teaching, it shows the kids that hey, your teachers are also learning. They see us stop and have a conversation that they're not a part of, and they're just kind of like, okay. But they see that we're also learning, so they … I don't know, I think they kind of respect that they're not just the students, but we're also the students." A few teachers mentioned anxiety about conducting a lesson with peers but also indicated that nervousness was ameliorated by a supporting community. "But it's good. I mean, we have a really good community here, so it's not like they're being judgy or anything like that, but I just get nervous sometimes." Counterbalancing anxiety was appreciation of the potential benefits of having peers in the classroom while conducting a lesson (expressed explicitly by 73%). For example, one teacher mentioned, "Oh, the experience of having the team in the classroom is so valuable. You really do get really good feedback, and a lot of times I know that people come back with videos, like, because I can't be everywhere and so it's really nice to have a lot of people invested and interested in what your class is doing and to give you feedback on these kids. I love that. It's my favorite thing, really." Another said, "It's just really difficult to do all of those things on your own, and so being together and having a group of kids and having every kid accounted for in that class to see exactly where every student is, I mean, it's so worth it."

Teachers' views of mathematical investigations

Eighty percent of the teachers attributed *increased understanding of the mathematics of measurement* to mathematical investigations that they conducted during the course of professional development. A representative teacher response was "I mean, I'm learning something every time we have a session, like when we did the Zoom about the multiplicative comparison, that was a totally new way of thinking about it for me." Another recalled, "And so, I remember in particular this one time when we were working on, like, a multiplication sentence and where to put parentheses and, like, why they were there and what this visually represents." A third suggested that both children's and teachers' investigations are important to learning: "But the point being that the teachers are having to struggle through just like the kids do, so that we're learning the mathematics similarly to the way kids would. Any time that we're doing something where we're experiencing it, that's good. We're learning." Most (87%) indicated that mathematical investigation tended to *impact teaching practices* by expanding the mathematical meanings of measure. For example, one teacher recounted, "I'm realizing, oh, that's why that makes sense. Then I can apply that in my classroom and I can let them learn about measurement through more of the discovery and conversations and hands-on, rather than me just saying, 'This is how you do it. Don't have any gaps' …. I think that's the biggest thing that I've gotten out of it for the younger kids is that they're really embedding that why and how." Another proposed that she could now help children understand arithmetic operations conceptually: "This is what is really happening instead of 'Here's what you multiply or here's what you add.'" Several of the teachers (40%) recalled how the *collective nature of mathematical investigations* contributed to the value of their experience. For example, one observed, "... it's really nice to have other people there to help you think and see things differently," while another suggested, "So just being able to sit and work out the math with other teachers whenever you're stumped, or drawing it out. You can see it in your head, but seeing their thinking and comparing our thinking is very productive."

Teachers' views of communal critique

As one of the teachers explained, "The main purpose of our after-school meetings is to continue to develop a deeper and richer understanding of what we're teaching ... and how students are understanding. And that involves a variety of things." The forms of activity cited most often were continued debriefing of learning labs (73%), conversations about construct validity (67%), and collective comparison of student learning within and across grades (93%). Debriefing of learning labs continued to center on student thinking and next steps: "how the lesson went and what we could do better in the future, and also where that group of students needs to go next and what they need to focus on, so we collaborate a lot on that." Conversations about construct validity were afforded by sharing records of student learning submitted as evidence of student thinking characteristic of a sublevel of a construct. As one teacher recounted, "We would go back and we would look at videos of the kids and finding evidence of the kids being in that certain construct. We would listen back and we would have to kind of point out, like, what the kids said or what the kid did to prove that they are on that certain construct. And that was always really helpful." Another suggested that this form of sharing also promoted collective accountability and broader perspectives: "Having that accountability, but also just having a group of people that can analyze that with you and strengthen your confidence in a way and also just to even see, well, maybe they can see something that you don't always see."

Teachers often mentioned that cross-grade comparisons tended to illuminate development. For example, one recounted, "Then we do some vertical alignment sometimes, where you might take something where, like fractions, parts of a unit, and we talk about how's that going to look in kindergarten, first, second, third, fourth, fifth. So, you're able to see that progression." Some teachers also mentioned that images of development had implications for their ongoing practice: "Like, me, as a first-grade teacher, when I'm looking at where some of the second-grade teachers' students are, it puts in perspective what's in between what I'm doing and what they're doing in second grade, and what can we do between now and then to fill that in." Teachers also revealed that within-grade comparisons of class-level progress on a construct appeared to promote curiosity and professional exchange. For example, one teacher explained, "Where you feel safe and you feel comfortable to recognize, oh well, kind of that ego blow of your kids are doing a lot better than my kids. Why?" In a similar vein, another proposed, "It lets me know, well, is my class kind of behind? Am I right there with them? What do I need to be doing to get them there? And then, if I see another teacher, their kids are farther along, I can just be, like, what are you doing that I'm not doing? How do I get them there? Then I can ask them for ideas on how I can fix that." Teachers also noted the value of both between-class and within-class TOTs visualizations for monitoring progress in learning. For example, "I think the star chart [a view of observations that characterizes the performance of individual students with respect to a construct] is helpful to kind of get just a glance of where the kids are." The display of students' performance by construct level helped some teachers focus on more equitable sampling of student thinking. For example: "I used it to help me just track, like, who have I talked to and who do I need to talk to? Like, I've talked to Sally 15 times, but I've only talked to José twice. So I know I need to really hone in with him and dig in and really spend some time with him so that I understand for sure. I may know really well where Sally is, but do I really know where José is?"

Changes in Teachers' Construct-Centered Judgments About Students' Ways of Thinking

Teachers used the teacher observation tools (TOTs) to record instances of student activity and work products as evidence of how students were thinking about core concepts of measure. Either in the moment during instruction or afterward, each performance was associated with a sublevel of a construct that reflected the teacher's judgment about the quality of student thinking that the performance revealed. We expected that as teachers repeatedly reviewed and redescribed student activities and products by assigning them to appropriate levels of a learning construct, teachers' perspectives—on both students and the constructs—would become progressively sharper and better aligned over time. In other words, a premise of the professional development aspect of the project was that the shared language of learning articulated by the constructs would support teachers in coming to observe student learning through this lens.

To test this premise, researchers conducted an audit of teacher judgments during the first, second, third, and fourth years of the project to ascertain the extent to which teachers' assignments of student thinking at particular levels and sublevels of the construct were warranted by supporting evidence, from the stance of a construct-conscious auditor. Evidence, which was recorded in

TOTs along with the teacher judgments, included photos, annotations to videos, the videos themselves, and teacher notes. Occasionally, a teacher included a judgment (that is, a sublevel assignment) but had insufficient time to warrant her inference about student performance by indicating the evidence that informed her judgment. If so, the teacher judgment might be valid, but its validity was not accessible to an auditor.

Because the length construct was focal across the entire K-5 spectrum (other domains, like volume and area, were only prominent at the later grades), we chose judgments about length as our window to evaluating teachers' construct-centered inferences of student learning. Records of observations in TOTs were used as the basis of the analysis. An auditor used the evidence that the teacher associated with each observation to independently make an assignment to a sublevel of a construct. When the auditor and teacher assignments coincided, the judgment was coded as "warrant traceable." Traceable warrants were further characterized by the auditor as either strongly supported or partially supported by the evidence artifacts. When the auditor and teacher judgments did not agree, because the auditor made an assignment to a different sublevel or level of the construct, the judgment was considered an instance of a "mismatch." When an auditor could not ascertain sufficient evidence to make an inference about a sublevel or level, the observation was characterized as "unclear warrant."

A second auditor coded a random sample of 20% of the observations assigned to the length construct. Exact agreement between the second and primary auditor was 76%. Percent agreement at the level of the construct (that is, the major levels, as opposed to the more detailed sublevels) was 85%. The two auditors reviewed their disagreements and resolved some minor ambiguities in the coding process. Subsequently, the second auditor coded another random sample of 20% of observations assigned to the length construct. On this occasion, exact agreement at the sublevels was 81%, and percent agreement at the major levels of the construct was 92%.

Teacher and auditor assignments of student to construct were compared for every teacher observation. Because a single observation could contain multiple teacher judgments (e.g., assignments to multiple sublevels of a construct), the contribution of each teacher judgment to the observation was normed to reflect the proportion of codes of each type, so that the contributions of all teacher observations were accounted for approximately equally. For example, consider an observation that required four teacher judgments about one or more students' thinking. Two of these were coded as warrant traceable, the third as a mismatch, and the fourth as an unclear warrant. Collectively, these judgments counted as a single observation, contributing 2 out of 4 observations to warrant traceable, 1 out of 4 observations to mismatch, and 1 out of 4 observations to unclear warrant.

TABLE 9.2 Teachers' Agreement with Auditor About Evidence
of Student Thinking

Audit results	Year 1	Year 2	Year 3	Year 4
Warrant traceable				
Strongly supported	32%	66%	70%	75%
Partially supported	6%	6%	6%	4%
Mismatch				
Within level	31%	2%	7%	15%
Different level	5%	5%	5%	2%
Unclear warrant	26%	21%	12%	4%

Table 9.2 displays the audit results by year of observation. The primary trend is a notable acceleration of auditor-teacher agreement. Overall, there was an increase in generation of evidence that was associated with a construct, and by the third and fourth years, there was a substantial decrease in judgments that could not be independently verified (i.e., those for which evidence was not available). In the fourth year, there was an unexpected increase in auditor-teacher mismatches within level, but most of these were accounted for by minor modifications of the length construct during the fourth year. For example, teachers proposed, and we adopted, an expansion of the second level of the construct so that teachers could distinguish between students who could explain how and why tiling units contributed to consistent measure and students for whom teachers could only observe the activity of tiling. However, during the fourth year teachers tended to assign both types of student performance to the same sublevel (i.e., to "tiles unit" instead of to either "tiles unit" or "tiles unit and explains why"), which accounted for much of the teacher-auditor mismatch within level during the fourth year.

Professional Vision as a Fulcrum for Learning Progression

Ongoing work, both our own and research conducted by others, is supporting a growing recognition in the field that learning progressions should not be regarded as static constructs that can be simply verified or falsified in a single "critical experiment" (see, for example, Confrey & Maloney, 2014; Duncan & Hmelo-Silver, 2009; Duschl, Maeng, & Sezen, 2011; Furtak, Morrison, & Kroog, 2014; Lehrer et al., 2014; Lehrer & Schauble, 2015). Rather than simply providing a series of developmental benchmarks that students meet or fail to meet, a learning progression can function as the core for what Lakatos termed a program of research, marked by continued inquiry, contest, and revision. In this chapter, we emphasize that learning progressions need not simply be ways to classify students; they can be engines of growth for communities of educators. We have explained how elements of the learning progression articulated in the preceding chapters supported teachers as they developed and exercised a professional vision of teaching and learning for student learning about measure. This kind of professional vision is an essential fulcrum for the realization of the possibilities of a learning progression, because all learning progressions presume particular conditions of instruction, even if those conditions are not always articulated.

Professional vision in the Mallard and Sleeve schools was diffracted through a series of conceptual tools that teachers employed to weave instruction and assessment. A central tool was the set of construct maps that collectively portray a midlevel view of the growth and development of core concepts of measure in length, area, angle, and volume. These core concepts and the developmental pathways described by the constructs are described in detail in earlier chapters in this volume. By a "midlevel view" we mean descriptions of students' ways of thinking that are more specifically tuned to learning of particular domain concepts than to the elements of a general learning theory and yet are at a higher level of description than the fine-grained strategies and production models that occupy cognitive psychologists. Midlevel descriptions of student thinking are those recognizable in students' talk and production in the classroom. Teacher noticings of students' ways of thinking were curated by teacher inquiry during classroom learning labs, which, in turn, set the stage for the development of a more responsive pedagogy among participating teachers. A construct observation tool (TOTs) brokered the use of constructs for systematically documenting the conceptual progress of individuals and groups of students. By employing TOTs representations, teachers could visualize classroom and school-level variability in student achievement. They were also more likely to notice when they had scarce evidence about particular students (e.g., "quiet" students), and as a result, TOTs also provoked teachers' efforts to support broader student participation. As teachers accumulated evidence of student learning over time, the extent of individual and collective student conceptual progress along a construct became increasingly visible.

At a community level, the constructs and TOTs allowed for within- and between-grade comparisons, so that students' progress could be represented across multiple classes within a grade and between grades. This capability sparked a broader view of school-wide progress as teachers became increasingly aware of how conceptual development at one grade set the stage for continued progress at later grades. We also instituted a communal practice of auditing teachers' evidence for construct assignment. Teachers regularly described their students' progress, presented evidence in support, and participated in conversations about whether other members of the community would make similar judgments, given the evidence presented. This activity promoted shared standards of evidence, so that, over time, teachers increasingly documented their judgments of student learning with an eye toward prospective peer critique, a form of standardization of measurement.

As teachers participated in video-stimulated interviews, much of what they noticed in instances of other teachers' practices in the videos was reflected in their accounts of their own practice. Moreover, transitions in what teachers noticed about video episodes of instruction were also echoed in their self-report of transitions in their teaching practices. Collectively, we took these transitions as signals of the development of a community of practice that was beginning to share a common way of viewing instruction and a common language for describing student learning and teacher intentions. This kind of shared community is a necessary precondition for establishing meaningful assessment that can foster responsive teaching.

References

Alonzo, A. C., & Elby, A. (2019). Beyond empirical adequacy: Learning progressions as models and their value for teachers. *Cognition and Instruction, 37*(1), 1–37.

Ball, D. L. (1993). With an eye on the mathematical horizon: Dilemmas of teaching elementary school mathematics. *Elementary School Journal, 93*(4), 373–397.

Ball, D. L., & Cohen, D. K. (1999). Developing practice, developing practitioners: Toward a practice-based theory of professional education. In L. Darling-Hammond & G. Sykes (Eds.), *Teaching as a learning profession: Handbook of policy and practice* (pp. 3–32). San Francisco, CA: Jossey-Bass.

Black, P., & Wiliam, D. (1998). Assessment and classroom learning. *Assessment in Education: Principles, Policy, & Practice, 5*(1), 7–74.

Black, P., & Wiliam, D. (2009). Developing the theory of formative assessment. *Educational Assessment, Evaluation, and Accountability, 21*(1), 5–31.

Confrey, J., & Maloney, A. (2014). Linking standards and learning trajectories. Boundary objects and representations. In A. P. Maloney, J. Confrey, & K. H. Nguyen (Eds.), *Learning over time. Learning trajectories for mathematics education* (pp. 125–160). Charlotte, NC: Information Age Publishing.

Confrey, J., Maloney, A. P., Nguyen, K. H., & Rupp, A. A. (2014). Equipartitioning: A foundation for rational number reasoning. In A. P. Maloney, J. Confrey, & K. H. Nguyen (Eds.), *Learning over time. Learning trajectories for mathematics education* (pp. 61–96). Charlotte, NC: Information Age Publishing.

Copur-Gencturk, Y., Plowman, D., & Bai, H. (2019). Mathematics teachers' learning: Identifying key learning opportunities linked to teachers' knowledge growth. *American Educational Research Journal, 56*(5), 1590–1628.

Davis, E. A., & Krajcik, J. S. (2005). Designing educative curriculum materials to promote teacher learning. *Educational Researcher, 34*(3), 3–14.

Davis, E. A., Palincsar, A., Arias, A. M., Bismack, A. S., Marulis, L. M, & Iwashyna, S. K. (2014). Designing educative curriculum materials: A theoretically and empirically driven process. *Harvard Educational Review, 84*(1), 24–52.

Desimone, L. M. (2009). Improving impact studies of teachers' professional development: Toward better conceptualizations and measures. *Educational Researcher, 38*(3), 181–199.

Duncan, R. G., & Hmelo-Silver, C. D. (2009). Learning progressions: Aligning curriculum, instruction, and assessment. *Journal for Research in Science Teaching, 46*(6), 606–609.

Duschl, R., Maeng, S., & Sezen, A. (2011). Learning progressions and teaching sequences: A review and analysis. *Studies in Science Education, 47*(2), 123–182.

Ehrenfeld, N. (2022). Framing an ecological perspective on teacher professional development. *Educational Researcher.* https://doi.org/10.3102/0013189X221112113

Ford, M. J. (2010). Critique in academic disciplines and active learning of academic content. *Cambridge Journal of Education, 40*(3), 265–280.

Furtak, E. M., Morrison, D., & Kroog, H. (2014). Investigating the link between learning progressions and classroom assessment. *Science Education, 98*(4), 640–673.

Ghousseini, H., Kavanagh, S. S., Dutro, E., & Kazemi, E. (2021). The fourth wall of professional leaning and cultures of collaboration. *Educational Researcher, 51*(3), 216–222.

Gibbons, L., & Cobb, P. (2017). Focusing on teacher learning opportunities to identify potentially productive coaching activities. *Journal of Teacher Education, 68*(4), 411–425.

Goodwin, C. (2018). *Co-operative action.* New York, NY: Cambridge University Press.

Horn, I. S., Kane, B. D., & Wilson, J. (2015). Making sense of student performance data: Data use logics and mathematics teachers' learning opportunities. *American Educational Research Journal, 52*(2), 208–242.

Kazemi, E., Gibbons, L., Lewis, R., Fox, A., Hintz, A., Kelley-Petersen, M. … Balf, R. (2018). Math labs: Teachers, teacher educators, and school leaders learning together with and from their own students. *NCSM Journal of Mathematics Education Leadership, 19*(1), 23–36.

Knorr Cetina, K. (1999). *Epistemic cultures: How the sciences make knowledge.* Cambridge, MA: Harvard University Press.

Lehrer, R., Kim, M.-J., Ayers, E., & Wilson, M. (2014). Toward establishing a learning progression to support the development of statistical reasoning. In A. P. Maloney, J. Confrey, & K. H. Nguyen (Eds.), *Learning over time: Learning trajectories for mathematics education* (pp. 31–59). Charlotte, NC: Information Age Publishing.

Lehrer, R., & Schauble, L. (2015). Learning progressions: The whole world is not a stage. *Invited Commentary, Science Education, 99*(3), 432–437.

Mesiti, C., Artigue, M., Hollingsworth, H., Cao, Y., & Clarke, D. (2021). *Teachers talking about their classrooms. Learning from the professional lexicons of mathematics teachers around the world.* New York, NY: Routledge.

Rotman, B. (1987). *Signifying nothing. The semiotics of zero.* Stanford, CA: Stanford University Press.

Samara, J., & Clements, D. H. (2009). *Early childhood mathematics education research: Learning trajectories for young children.* New York, NY: Routledge.

Sherin, M., Jacobs, V. R., & Philipp, R. A. (Eds.). (2010). *Mathematics teaching noticing. Seeing through teachers' eyes.* New York, NY: Routledge.

Strauss, A., & Corbin, J. M. (1990). *Basics of qualitative research: Grounded theory procedures and techniques.* Thousand Oaks, CA: Sage Publications, Inc.

Wilson, M. (2005). *Constructing measures. An item response modeling approach.* Mahwah, NJ: Lawrence Erlbaum Associates.

Wilson, M., & Lehrer, R. (2021). Improving learning: Using a learning progression to coordinate instruction and assessment. *Frontiers in Education. The Use of Organized Learning Models in Assessment, 6*:654212. https://doi.org/10.3389/feduc.2021.654212

10
MEASURES AND MODELS IN ELEMENTARY SCIENCE

It is tempting to think of spatial properties, which intuitively seem accessible to direct perception, as being self-evident. Yet, as we have argued, measuring, even of spatial extent, is inherently a constructive practice. It entails identifying attributes, an activity that demands focusing on relevant information while provisionally ignoring other available information, an aesthetic that must be cultivated. Developing measures further necessitates deploying and coordinating additional acts of the imagination, such as partitioning magnitudes of relevant attributes into units that are defined in relation to a scale, plus translating and accumulating the units to construct a measure. These acts of imagination have their counterparts in material means, such as specialized measurement tools, and are associated with intersubjectively agreed-upon ways of employing those tools. In short, the analysis of spatial domains in earlier chapters reveals that measurement requires considerable reconceiving of spatial magnitudes.

Although measurement is usually first introduced to students in their mathematics classes, it also plays a central role in other school domains, especially science. To achieve an empirical grasp on how natural systems work, those systems must be measured. Yet, as in mathematics instruction, measurement in science class is often restricted to learning to use tools, such as balances and thermometers, along with learning about the properties and commensurability of different measurement scales (Smith, Males, Dietiker, Lee, & Mosier, 2013; Thompson & Preston, 2004). This restricted vision is problematic, as the constructive aspect of measurement is, if anything, even more challenging in science than in mathematics, because it is so inextricably tied to people's conceptual models of the situation in which the measurements are recorded (e.g., Cannaday, Vincent-Ruz, Chung, & Schunn, 2019).

DOI: 10.4324/9781003287476-10

Science education researchers have paid considerable attention to the importance of evidence—often in the form of measures that are cumulated and displayed—in holding models of natural systems to account (e.g., Klahr, Fay, & Dunbar, 1993; Kuhn, Amsel, & O'Loughlin, 1988; Schauble, Glaser, Duschl, Schulze, & John, 1995). By *model* we refer to a system consisting of attributes and relations among these attributes that explains some aspect of a natural phenomenon (Lehrer & Schauble, 2004; Nersessian, 2008). Modeling always depends on decisions about relevant attributes and their relations, but how measures participate in the formulation of attributes is usually neglected in educational research (Manz, Lehrer, & Schauble, 2020). However, research in the history, sociology, and philosophy of science clarifies how frequently we come to understand what an attribute means only as we settle on how it can be measured. That process sometimes continues over an extended period—indeed, sometimes over many years, as exemplified by the gradual development of measures of heat and temperature (e.g., Chang, 2004). The history of science provides accounts of the iterative bootstrapping that occurs as scientists struggle, on the one hand, to identify a theoretically relevant attribute of a system and, on the other, to find an informative and reliable way to instantiate it in empirical phenomena so that it can then be measured (Pickering, 1995). How an attribute is conceived of influences how it is measured, and the process of developing instruments and measures, in turn, increasingly brings attributes of a phenomenon into focus, resulting in a cycle of ongoing development and revision.

With this analysis in mind, we look to the sciences, which feature contexts that extend the reach and grasp of measure as students investigate the workings of natural systems. Characterizing natural systems creates the need to develop new kinds of measured quantities, including those that have very broad application across scientific domains, such as time, temperature, and weight, as well as those whose meaning applies within narrower subdomains, such as ecological carrying capacity, air or water pressure, and electrical conductance. Getting a grip on natural phenomena also motivates the development of new means and methods of measure, so the productive tension between imagination and material means that arises in spatial measure is broadened and made more general. For example, how does one arrange conditions to measure growth of an organism or to indicate what stays the same when the mass or volume of a material changes? Important properties of natural systems are often made accessible only by creating new relations among measures, such as density to characterize material kind and rate to characterize change over time in growing organisms or in the movement of objects. Constructing relations among measures can also introduce children to new mathematical systems, such as Cartesian coordinates that align multiple rulers in space to indicate location, polar coordinates that describe location by relating angle and length, and distributions of measured values visualized by repurposing Cartesian axes to display along the scale of measured values the frequency with which a value of a measurement recurs (e.g., a dot plot). As we will explain, coordinates and distribution exemplify new realms of mathematics that come into focus when children are positioned as measurers of attributes of natural phenomena.

Constructing measures also typically enhances the course of investigation and inquiry. For example, in a sixth-grade classroom, students investigating the ecology of local retention ponds designed microcosms (gallon jars with substrate, aquatic plants, and animals as well as inadvertent inhabitants, such as algae and bacteria) that replicated selected aspects of the ecosystem to answer questions that they and their classmates posed (Lehrer, Schauble, & Lucas, 2008). As the children worked with their microcosms, their measures of microcosm functioning often inspired refinement of those questions in ways that made it more feasible to generate relevant evidence. For instance, an initial question, "What are the effects of fish and frogs?", indicated an interest in examining impacts of these animals on the microcosm ecosystem. The question was subsequently refined as students focused on how to measure these impacts: "Who consumes more dissolved oxygen?" This reframing provided a practical avenue for examining "effects" and also suggested other potential effects due to additional aspects of water chemistry.

To illustrate the importance of measure in scientific inquiry, this chapter presents two separate but related developmental accounts, both illustrating the centrality of measure in scientific reasoning. First, we look through the lens of content domain, showing how the "same" content can be understood in increasing depth as children's appreciation of measure expands. Then we shift our focus to investigate the development of students' grasp of a new kind of measure, one critical in the sciences—statistics, or measures of properties of distributions.

The unifying theme of the first section is how measure supports changes in student thinking about the growth of organisms. The purpose is to illustrate how transitions in students' grasp of measurement across the elementary grades open the possibility to pursue progressively more sophisticated forms of investigation within the same content area. Students' developing repertoire of measures was deployed to conduct investigations first about the growth of individual organisms and then about the growth of populations of organisms. The shift from individual organisms to populations required a concomitant turn to focus on distribution, a form of thinking that is fundamental to scientific practice. Distributional thinking, in turn, required new kinds of measures.

The second major section picks up this theme of distributional thinking and explains how its foundations were laid across the grades. The account illustrates our earlier claim that new domains of content bring new realms and properties of measure into focus, especially variability. As in the previous section, we begin by describing the youngest students' initial attempts to interpret variability and then recount how descriptions of distributions became more sophisticated and precise in subsequent grades as children's related mathematical repertoire also expanded. The section culminates with an account of sixth graders' work to generate and test models of distributions. Throughout both sections, we illustrate how an increasing grasp of measures broadened students' ability to formulate and pursue questions of their own interest.

Characterizing Growth

Explorations of how organisms grow and change during their life span are prominent in most elementary science instruction. Qualities of measure can contribute to the nature of claims and evidence that children bring to bear on questions of interest during these investigations. To provide a sketch of measure's developmental potential in the sciences, we summarize results from studies more extensively described elsewhere (the full studies are cited as they are mentioned). Our purpose is to illuminate the productive interplay between measure and the conduct of investigation.

Describing Change by Determining Differences in Quantity

Children in a first-grade class investigated the growth of two different species of amaryllis plant over several weeks to ascertain whether the plants grew "better" when planted in soil or in water (Lehrer & Schauble, 2000). To prepare for this investigation, children observed the growth of a single species of amaryllis and drew what they observed at two timepoints: shortly after the bud first appeared and again after it flowered. Drawings were situated within a broader investigation of kinds and functions of scientific representations, intended to help students come to see representations as highlighting some parts of a natural system, even as other aspects of the system receded to the background.

Many children struggled with how to describe change, and the teacher asked the class whether the drawing could be used to decide *how much* the plant had grown. As students considered this question, someone proposed measuring, and a conversation ensued about precisely what should be measured. Should height include only the stems or also the flowers? As Chapter 3's account of the comparison of pumpkins revealed, attributes identified as candidates for measure must be defined, along with the proposed method of measure. As children debated how to identify the "actual height" of a plant, the author of one drawing sketched in "strips" that were intended to stand in for the heights of the plant on both days of observation. The drawn strips aligned with the tops of the flowers at the greatest distance from the base of the plant. These notations inspired the related idea of cutting paper strips to directly compare plant heights over time, a strategy that children next pursued as they grew new rounds of plants to investigate how different species grew in different media. Figure 10.1 depicts changes in the heights of two species of amaryllis and two narcissus plants, some grown in soil and some in water.

The hard-to-read text on the strips in Figure 10.1 displays the date on which the plant's measure was recorded. However, because children measured the plants only on days when it was convenient to do so, the intervals of measure were not uniform, and time was not represented explicitly

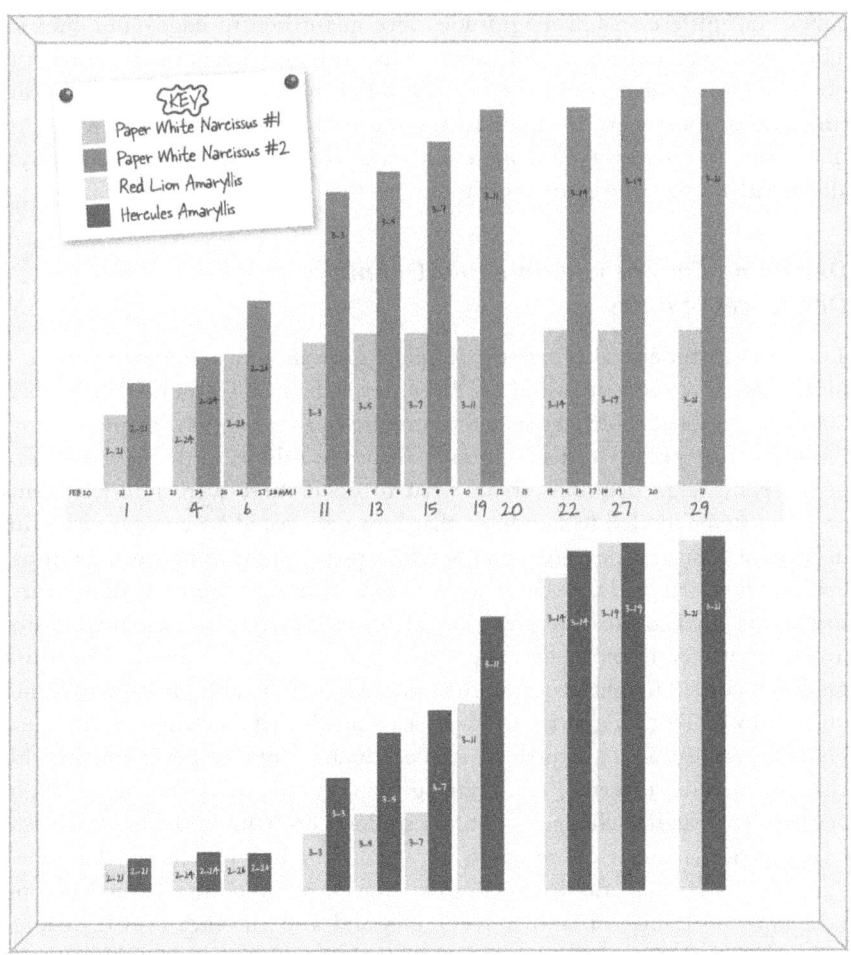

FIGURE 10.1 Display representing heights of flowering bulbs grown in soil and in water

as a dimension (see Levin, Israeli, & Darom, 1978, and Thomas, Clarke, McDonough, & Clarkson, 2016, concerning children's conception of time as focusing on succession of events rather than duration). For these reasons, the display only approximately reflects the pattern of plant growth, but it did provide students with opportunities to compare differences in heights between species and conditions on the same days, as well as within species at earlier and later points in the life cycle. Nonetheless, these direct comparisons were not responsive to the teacher's original comparative question about "how much more or less." Accordingly, she next asked students to propose how the length of each strip could be measured. After exploring several candidates for a unit, children settled on the edge of a Unifix™ cube, a familiar tool in plentiful supply. Strip lengths were tiled with Unifix™ cubes, and differences in magnitude were quantified in edge-cube metric. There was some debate, as well, about what to report when the measure was not a whole number; the issue was amplified in future lessons, but at this time partial units were confined to approximations of "one-half." With this innovation of quantity, students could now compare species in new ways, albeit still at only two points in time.

Describing Change with Intensive Quantity: Differences in Rates

Third-grade students also investigated plant growth with a model organism—in this case a Wisconsin Fast Plant™ (Lehrer, Schauble, Carpenter, & Penner, 2000), a species of cabbage selected to mature in about 40 days from planting. As in the first grade, students initially focused on how to define the height of the plant (e.g., decisions about what to do if there were multiple stems or if the stems tended to bend), so that measures could be compared within individual plants as they grew and between plants at the same times of measure. By the third grade, students were already familiar with units of measure, so they decided to measure plant heights in millimeters because these units, unlike centimeters or inches, supported whole-number counts. The third graders recorded height, day, and time of measure in a table, along with other observations. To characterize growth, students described change with a new kind of quantity, a rate, expressed as a coordination of changes in height with changes in time, to arrive at amount grown, expressed as change in height per day. To visualize change over time, students drew upon the mathematical resource of Cartesian coordinates, a system they had previously developed during their mathematics instruction as a way of relating the measures of long sides and short sides of similar rectangles and also the measures of heights and circumferences of similar cylinders (Lehrer, Strom, & Confrey, 2002). Now, however, only one dimension (the height) was measured, while days of

Height of Round Two Fast Plants
(6 pellets fertilizer)

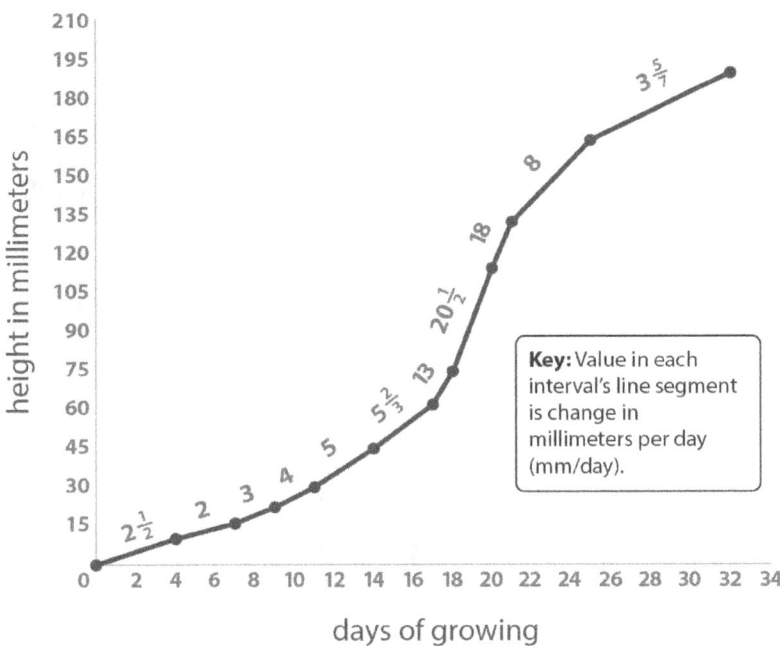

FIGURE 10.2   Visualizing change of plant height in a Cartesian coordinate system

observation were translated into ruled time. Even though the plants were not measured at constant intervals (weekends intervened), on the graph days of measure were plotted on a consistent scale. Changes in the rate of growth of single plants were represented by line segments that spanned different intervals in the Cartesian system, as illustrated in Figure 10.2. As children looked across similarly designed coordinate graphs of different individual plants, they noticed that most of the plant growth curves appeared to take a form that they described as an "S-shape." This shape corresponded roughly to an initial period of growth from seed, an acceleration associated with the production of flowers, and a subsequent deceleration as the plant produced seeds and tended toward senescence.

Following conversation about the general form of the growth curve, students superimposed the heights of many plants, measured on common days of observation, on the same graph. Doing so provoked a new question: Could an S-shape be drawn to represent the entire collection of plants, even if some points on this graph did not intersect with any particular plant?

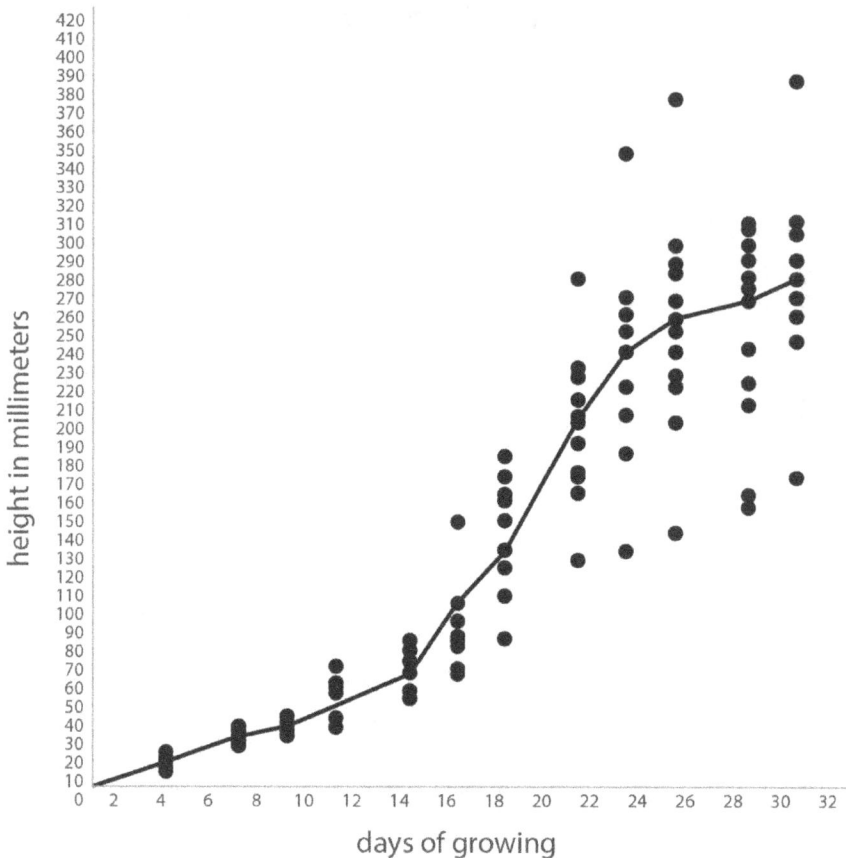

FIGURE 10.3 "Typical" height inscribed for a collection of plants

This question was resolved by approximating the height at each day of growth as the mid-range of the collection of points on that day of growth and then re-inscribing the curve to describe plant growth in general, as shown in Figure 10.3. Approximation of this kind is a hallmark of modeling. The teacher suggested, too, that students investigate changes in the curve that would result from changing the interval size. These explorations helped children appreciate that the shape of the curve also depended on interval and that other intervals could be imagined, even if they were pragmatically difficult to accomplish. For example, children imagined how the curve might change if they could measure plant height every day, every half-day, or even every hour.

During a second cycle of investigation, the scope of children's inquiry expanded to include new attributes of the plant and new comparisons, such as root vs. shoot growth and changes in the volume of the plant canopy over time.

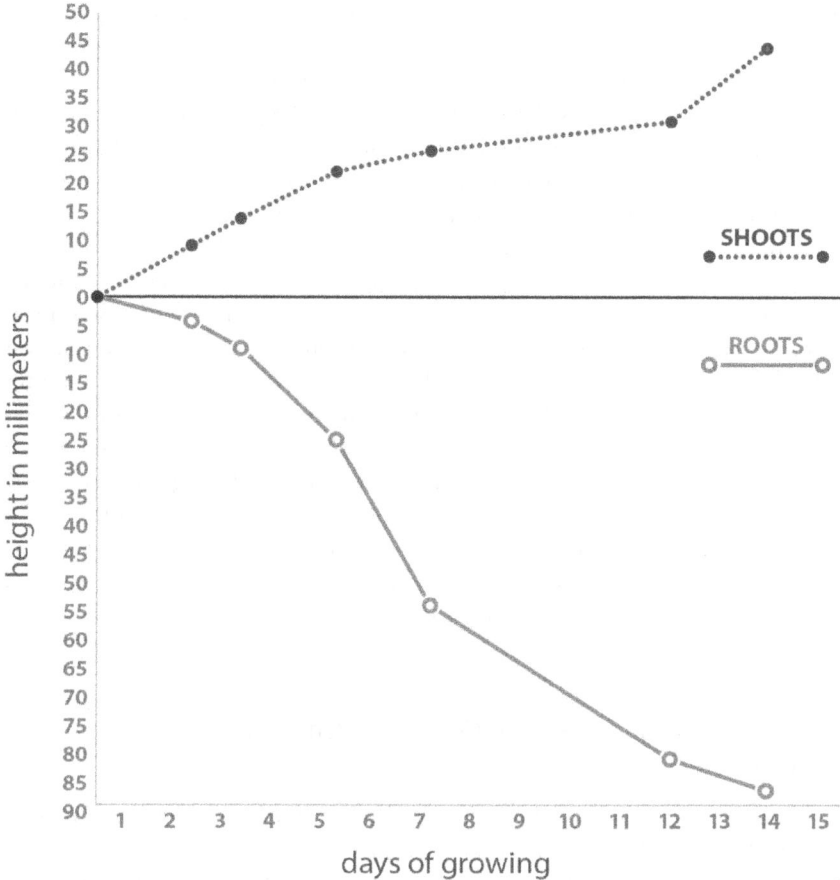

FIGURE 10.4 Expansion of coordinate system to compare root and shoot growth

To compare root and shoot growth, students expanded the Cartesian system as depicted in Figure 10.4, an innovation that allowed root and shoot growth to be shown in the same representation. Students noticed that the root growth showed earlier acceleration than the shoot growth, an observation that led to a discussion of the role of the root system in the survival of the plant. Students also noticed that, in spite of differences in their rates, both root and shoot growth exhibited the now-familiar S-shape.

Students also examined changes in the volume of the canopy of individual Fast Plants™ by approximating their volume, using either a prism with a square base or a cylinder. Modeling volume was based on two measures of growth: plant height and plant "width," defined as the maximum length from a leaf tip to its counterpart leaf tip on the opposite side of the stem. Putting these measures to use in the construction of the prism was straightforward, but some students objected that the prism was too imprecise an approximation of the volume of a plant's canopy; they preferred cylinders. Thinking of a cylinder as a "wrapped rectangle" meant that one edge of the rectangle was defined by the plant's height. But the other edge would have to be the circumference of the cylinder. Yet the only measure students had was the plant width, which corresponded to the diameter of the cylinder. Encouraged by the teacher, students constructed several circles with diameters of different width and traced the outline of the circumference with string. Inspecting these different instances of circle, students noticed a new invariant relationship: The circumference was always about $3\frac{1}{5}$ times as long as the diameter. Armed with this new, unexpected relation, students went on to construct cylinders and investigate changing volumes of individual plants on different days of growth. These investigations were grounded in students' previous invention of thinking about the volume of a cylinder dynamically, as the area of the based moved through its height (see Chapter 5, this volume, and Lehrer, 2000; Knapp & Lehrer, 2005). With these three-dimensional models, students were able to make new claims about how the volume of the canopy changed and were also able to align this new model with the previous S-shaped growth curve that characterized root and shoot growth.

Describing Change in Population with Measures of Distribution

As in earlier grades, fifth-grade students also investigated the nature of growth of Fast Plants™, but at this grade the focus was on the population level. Students developed experiments to address the effects of different levels of fertilizer and light on plant growth (Lehrer & Schauble, 2004). Conducting these experiments and shifting their attention to collective patterns in growth required that students organize a collection of measures as a distribution. Distributional thinking rests on a complex set of related concepts, including developing a variability-generating process that often includes random components (Saldanha & Thompson, 2002), coming to see a collection of measured outcomes as structured (e.g., different densities of outcomes), and understanding that properties of a distribution can be measured (via statistics) to further scientific investigation. Each of these ideas must be constructed with students.

To initiate a shift toward population and distribution, the teacher referred students to their collection of measures of height at the midpoint of the plants' life cycle and asked them to invent a way of visualizing what they noticed about what was "typical" and how "spread out" the measures were. The teacher intended to rely on the variability of student-invented displays to highlight how different shapes or visual patterns in data emerge from the choices that designers make about what to "show and hide." Students' inventions were indeed highly variable. Some of them appeared to have been guided by a preference for constructing displays that were unique, rather than readily interpretable by others. For example, in Figure 10.5, intervals of 10 are demarked on the *y*-axis and units representing ones are on the *x*-axis. This kind of thinking is similar to the thinking that underlies some conventional displays, like stem-and-leaf plots, but during follow-up comparisons of displays students noted that the display made it difficult to discern any structure in the collection.

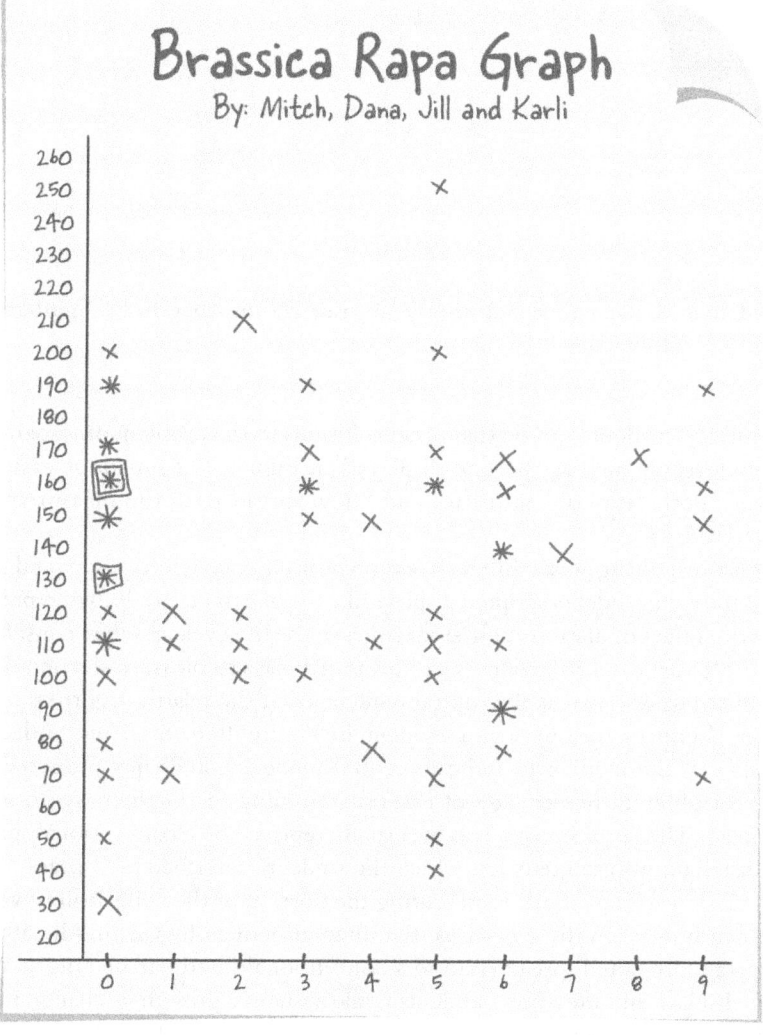

FIGURE 10.5 Student-invented display to visualize distribution via coordinates

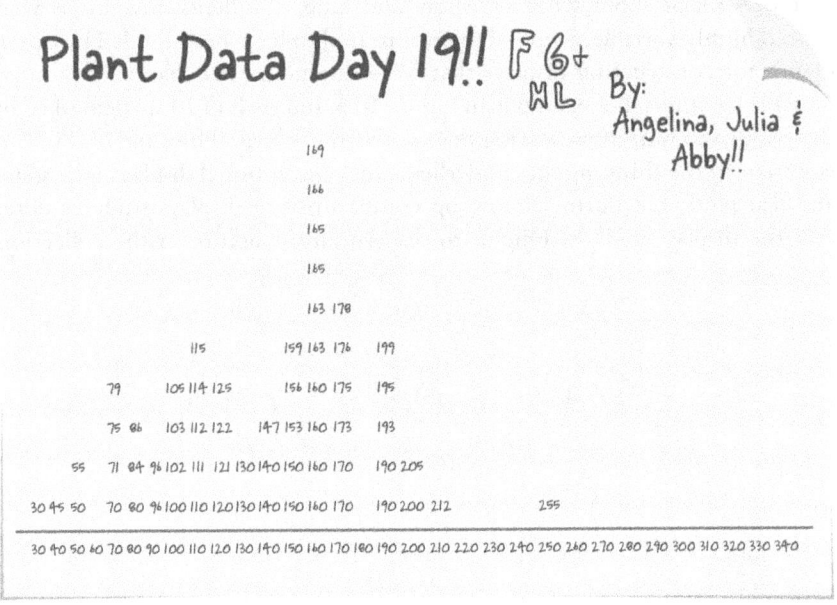

FIGURE 10.6 Student-invented display to visualize distribution by highlighting frequency

Ultimately, students agreed that other solutions to the problem of visualizing the collection, such as the one displayed in Figure 10.6, made it easier to identify both "typical" measures and "how spread out" the measurements tended to be.

After discussing what different invented displays tended to show and hide about the data, students adopted displays like the one in Figure 10.6 to represent growth of the population at different times in the life cycle. As they considered and compared displays, students also proposed ways to measure characteristics of a distribution, such as the central tendency and the relative tightness of the center clump of measurements evident in Figure 10.6. Students settled on statistics of the median to indicate central tendency, and they proposed the range of different percentages of cases surrounding the median as a measure of spread. The latter statistic was eventually replaced by a conventional split of the distribution into quartiles, which the students described as "hinges" with "doors" of varying length representing the portion of the collection of values between hinges. In these displays, the interval defined by the middle 50% of the data represented spread (akin to a conventional interquartile range). With these displays and measures, students could visualize growth as change in the shape of a distribution, from a positive skew early in the life cycle, due to the dominance of very low heights, to a more symmetric, "normal" distribution at the midpoint of the life cycle, and then to a negative skew late in the life cycle as plants approached senescence and reached their ultimate height. Changes in measures of a distribution's center and spread accompanied these changes in the shape of the distribution. This distributional view was a new way for students to visualize change.

During the conversations about display and measure of distribution, the teacher often asked students to predict what they would see if they "grew the plants again," an imagined repetition of the process (Saldanha & Thompson, 2002). What would they expect to see if they inspected qualities of a distribution at a particular day of growth? Some students predicted that measures in the center clump or measures close to those values would be more likely to recur, although other children took the stance that every height was equally likely to be observed again. To achieve a better grasp on this imagined process of regrowing, students next undertook sampling experiments to simulate what would happen if the populations of plants were grown repeatedly under identical conditions. They did so by compiling cards that recorded each measured height at the 19th day of growth (about the midpoint of the plants' life cycle), putting the cards into an envelope, and randomly drawing samples of different sizes. The resulting samples were then displayed as distributions, and the qualities of those distributions were discussed by the students. The point of this activity was to invoke the image of a random repeated process with a sampling distribution as a way of identifying the outcomes of those repetitions that were more and less likely to recur. In this context, students experienced opportunities to reason about variability arising from chance, both within a sample and from sample to sample. Students noted that, in general, the samples resembled the parent population, but central regions of the data seemed to most frequently reappear in the samples. Next students worked with a computer tool that mimicked what they had performed with physical materials (a capability now routinely supported in software like TinkerPlots™ and CoDAP™). These tools made it possible to rapidly produce many samples and to record statistics of center and spread. Students went on to simulate how medians tended to be distributed from sample to sample, just by chance, producing a new kind of distribution—the distribution of sample medians. This distribution provided an image of the variability one might expect to occur randomly.

With these images of random variability in mind, students contrasted samples of plants grown under low light or high fertilizer with a larger sample of plants grown under standard conditions. The teacher divided the class into small groups and challenged each to invent a method for comparing the distributions. Students generated many strategies for informing their decisions about the effects of light and fertilizer. These ranged from (1) comparing single values (most prominently the sample medians) and taking any difference between medians as evidence of an effect to (2) looking at the proportion of cases in each distribution that fell within set limits and (3) comparing the value of the median for the plants grown in lower light or with more fertilizer to the clump of median values expected by chance when plants were grown in standard conditions, according to their previously conducted simulation studies. This last strategy explicitly capitalized on chance variability, with median values that fell outside of the center clump of simulated medians taken as evidence of an effect. Follow-up interviews about student reasoning revealed that most of the students' inferences were guided by considering chance and then thinking of outcomes as legitimately different—and, hence, as caused—only if they were unlikely to be generated solely by chance. When we probed further about sources of chance variability under standard growing conditions, students tended to cite abiotic factors, such as chance variations in moisture, light, and fertilizer. Almost no student mentioned genetic recombination, suggesting that processes that generate natural variation may be opaque to elementary students. For this reason, in the next section we describe some alternative contexts for cultivating distributional thinking in which students can experience first-hand variability-generating processes. However, as we further elaborate, we do not advocate avoiding distribution when generating processes are less directly experienced, but instead seek to craft instruction where distributional thinking can be cultivated systematically throughout the elementary grades.

Cultivating Distributional Thinking

As we noted, distributional thinking refers to a network of related concepts, including appreciating a sample as generated by a long-term process that may include sources of chance variability and understanding that variability can be structured and measured. These conceptual foundations can be elaborated informally throughout elementary schooling.

Initial Steps

A good way to begin cultivating distributional reasoning is to expose even young children to opportunities to entertain and justify interpretations of differing outcomes of a process with which they have first-hand experience. These experiences should produce enough data so that deciding what the data mean is a legitimate problem, but not so much data that youngsters lose an intimate sense of connection between the data and the situation that produced them. For instance, first graders who built Lego™ cars of varying designs conducted a series of trial races on an inclined plane to determine which design features affected the speed of the cars (Lehrer, Schauble, Carpenter, & Penner, 2000). The children used a stopwatch to measure how long the cars took to traverse a standard-length racetrack. They ran 10 trials with each design, resulting in a small collection of measures for each car that represented the time it took that car to cross the finish line. At first, children were perplexed by the variability of the time measures from trial to trial—there were accusations that perhaps some participants were not operating the stopwatch correctly. To students' surprise, conducting additional trials with enhanced vigilance did not yield the expected homogeneity of results. Eventually, students concluded that it was impossible to entirely standardize the initiation and conclusion of time on the stopwatch, because it seemed impossible to control delays between the verbal order to release the car and the initiation of the stopwatch. Sometimes stopwatch operators responded quickly, while on other occasions they were slower. These individual differences resulted in measures that were sometimes "a little more or a little less." Students concluded that they should expect some variation but, faced with a small collection of measures, still wondered how to agree on a measure that could fairly represent the time that each car traveled.

This episode was the children's first attempt at trying to make meaning of multiple data points. Many of them did not, in fact, entertain the entire collection of values, but focused instead on individual values. Usually, they simply chose the fastest outcome to represent the speed of their favored car design, ignoring values that fell into the lower part of the collection of outcomes. A few children were drawn to values that appeared more than once in the data, reasoning that an outcome that occurred more than once was "probably right," regardless of its location within the collection (most children did not think to order the values and therefore did not attend to the location of the value they selected with respect to the locations of the rest of the outcomes they had recorded). No student proposed a value because it was located toward the center of the data points. This experience was only a modest first step on a long road toward working with data in the form of multiple outcomes. Educators' follow-up questions encouraged students to think of the car races as a series of repeated events that vary but, nonetheless, also show some structure: "What happens when you run a car again and again? Why? If we did it five more times, what would you most expect to see? What outcomes would surprise you?"

Although timing the descent of cars on a ramp had the merit of providing a first-hand experience of process, other forms of inquiry can also cultivate young children's appreciation of distribution. A different class of first-grade children compared the relative abundance of different species of insects living in the forest and prairie near their school (Lehrer & Schauble, 2012). After several rounds of field collection and conversation about the implications of different methods of collection (e.g., ground traps vs. close mesh nets), their teacher introduced a way to transform counts of different species into a

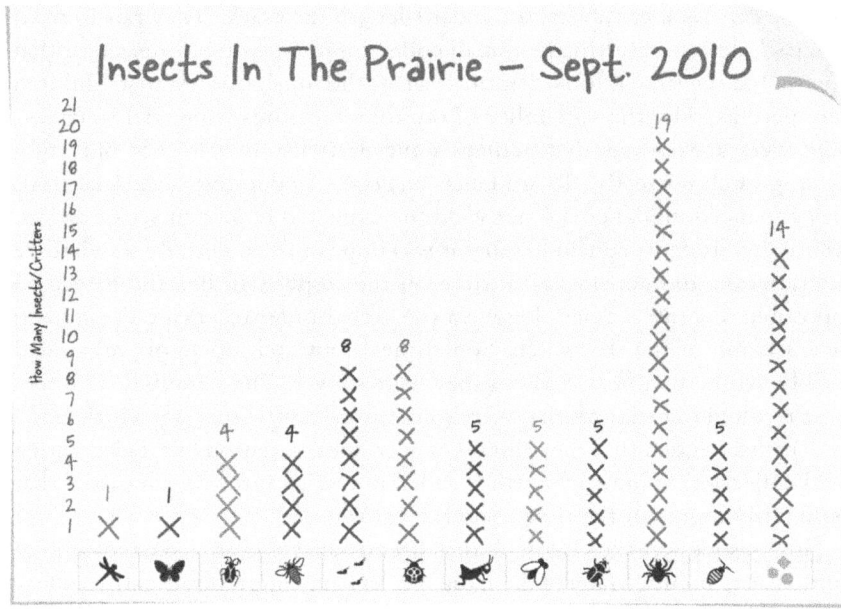

FIGURE 10.7 Distribution of data on relative abundance of insect species in the prairie

visualization, shown in Figure 10.7, that facilitated inference about the relative abundance of species at each place. Figure 10.7 was constructed by merging samples of prairie insects that were collected by pairs of children.

The visualization sparked conversation about the variability-generating process: Why were some species seemingly more abundant in each place (e.g., spiders, although not "insects"), and why were species evident in one place but not the other (e.g., bees in the prairie but not in the forest)? During these discussions, students evaluated each place as a source of resources to meet the needs of species there. For instance, both forest and prairie were rich sources of food for spiders, the prairie had more flowering plants with nectar for bees, and the forest had fallen tree limbs that provided shelter and food for sow bugs. These discussions promoted thinking about the distribution of species abundance as being due to ecosystem mechanisms framed as needs and resources for different species. The teacher also sought to promote the idea that the sample exhibited in the distribution was predictive of future samples. She did so by asking students to predict the relative abundance the class would be likely to observe if they collected insects again later in the week. Some students were convinced that the number of each insect species would be identical to what it was for the group collected earlier, but others suggested that sometimes they were just "lucky" to capture a species with low abundance, so the outcome would be "the same [relative counts] but different [not exactly the same count]."

Other children suggested that luck meant that any outcome was possible; they did not see the sample as predictive. This conversation was a fruitful starting point for considering samples as varying just by chance, although the meaning of chance was not well articulated. Nevertheless, even for young children, visualizing distribution in this context facilitated initial access to the poles of cause and chance in a data-generating process.

In subsequent grades, further opportunities to reason about distributions emerged as children encountered situations in which the number of data points increased sufficiently to result in a discernible shape of the data. By *shape* we mean visible patterns in a collection of data that emerge due to choices made by designers of data visualizations. Shape is imposed on data. Experiences with measurement that emphasize ideas about partitioning and fractional proportions are relevant to identifying and quantitatively characterizing segments of the distribution—and, eventually, to proposing measures of the distribution.

Figure 10.8 displays one of two distributions of bubbles that a third-grade class made with different "mystery" solutions—one colored blue and one pink. The purpose of the investigation was to decide whether one of the solutions created "better" bubbles than the other. Students proposed and then explored two senses of "better"—in one case the largest bubbles and in the other bubbles that lasted the longest before popping. Figure 10.8 displays results of trials on the outcome of bubble size, operationalized by popping each bubble on a

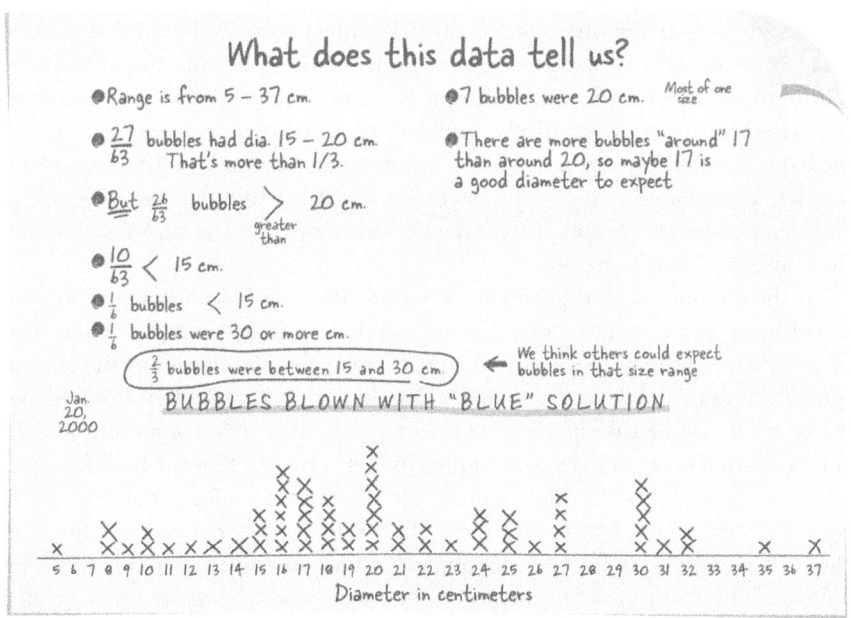

FIGURE 10.8 Partitioning of observed variability in bubble diameters

paper towel and measuring the diameter of the resulting wet spot. Each child blew three bubbles, popped them, carefully measured the diameters in centimeters, and recorded the results. In the Figure 10.8 display, the teacher has cumulated and plotted the diameters of all the bubbles, each represented as an X on a frequency display, with repeated values indicated by the height of a column. The display also shows notes recorded during the class discussion about the display.

The teacher asked what students would expect to see "if we blew the bubbles again." Main points that emerged from the discussion, summarized in the teacher's notations on the display, show that some students (like the earlier first-grade Lego™ car designers) initially focused on repeated values. They said they would expect to see bubbles of the size that most frequently appeared in their distribution: "Seven bubbles were 20 cm," so 20 cm was identified as the prediction for the "most of one size" that would be seen "if we did it again." However, other students objected that it was improbable that a new trial would produce an exact copy of the same measures. So, one of them proposed an interpretation based not on a single value but on "clumps" in the data—that is, dense batches of data that clustered around one or a few adjacent values. He argued, "There are more bubbles around 17 than around 20, so maybe 17 is a good diameter to expect." This student went on to acknowledge that although six of the bubbles had diameters of 20 cm and only five bubbles had a diameter of 17 cm, nonetheless the 17-cm bubbles were positioned within a large clump of adjacent measures, whereas the clump around 20 cm was less dense—it was surrounded, in the student's words, by "holes" in the data. Students next turned their attention to further defining this quality of "clumpiness" and whether there might be some way to quantify it. Students calculated the fraction of all the bubbles represented in different clumps of the data. This shift in attention from "numbers in a clump" to "fraction of all the data represented in a clump" occurred readily in this class because of the children's previous extended experiences with partitioning and fractions in the context of length measure.

As the comments on the left side of Figure 10.8 show, students next sought to delineate more precise cut-off points to define their "clumps." They first identified the intervals between 15 and 20 cm as the boundaries of the largest center clump of the data. Then they calculated the fraction of all the bubbles that were located either below that clump or above it. After some initial calculation, students concluded that approximately one-sixth of the bubbles were 20 cm or greater (at the high end of the distribution) and about one-sixth were less than 15 cm (at the low end). On this basis, they determined that the clump between 15 and 30 cm encompassed the remaining two-thirds of the bubbles. The third graders concluded they were comfortable predicting that this was the "clump" where most of the bubbles would probably occur "if we blew them again."

Students' varied ways of thinking about this distribution—greatest number of the same value (e.g., mode), "clumps" and "holes" in the data, center clump, fractions of the data represented in clumps, shape of the data expected if the trials were to be repeated—are forms of reasoning that we and other researchers have observed in young students' thinking about distributions across a variety of contexts (Cobb & McClain, 2004; Lehrer & Schauble, 2004; McClain & Cobb, 2001). They are useful foundations that support children's sense-making and can bootstrap their later understanding of more canonical ways of measuring characteristics of distributions with statistics.

Revisiting Cause and Chance in Generating Variability

Almost all of the first-grade students and a minority of the third graders did not initially predict that replications would vary somewhat from trial to trial, even though most of them eventually came to expect some variation just by chance. Understanding chance is usually challenging, because humans tend to think of causes as determining outcomes in a straightforward way. This bias toward determinacy is so well established that people tend to "see" cause-effect relationships even when the data suggest that they are not operating. In general, people tend to over-attribute evidence of causality (whether or not that is what the evidence suggests) and are much less inclined to mentally represent lack of causality between a potential precursor and an outcome (Kuhn, Amsel, & O'Loughlin, 1988; Schauble, 1990). For instance, when they are provided with data that show lack of covariation between episodes of a potential cause and a presumed effect, individuals sometimes focus instead on the positive correspondence between one or more precursor episodes and the outcome, ignoring instances when the presumed causal event occurs but the result does not (or when the presumed cause is absent but the outcome nonetheless occurs) (Kanari & Millar, 2004).

People are especially inclined to focus on cases that confirm their prior beliefs and to ignore disconfirming cases (Kuhn et al., 1988). For instance, suppose children believe that crayfish are more likely to live in the sections of a stream that are well shaded. Although data may show that crayfish live in both shaded and unshaded areas, a child may support his prior belief by citing only those instances when crayfish are found in the shade. An individual who selectively views the data in this way may miss the big picture, because chance reveals its structure only over a large enough number of trials and usually against a background of trial-to-trial variation. In these situations, people's biases may easily influence what they "see" and thereby what they conclude. Young people, for example, often presume that all animals are intentional, so the behavior of animals is perceived as being motivated by conscious choice. Against such a set of assumptions, the idea that some behavioral responses occur merely by chance seems counterintuitive and may be rejected or may not even be entertained. Moreover, the devices used to model chance situations (e.g., coinflips, spinners, tosses of a die) do not in any obvious way resemble the objects in the world whose behavior they are intended to represent, and as a result, children sometimes reject their legitimacy as tools for modeling behavior.

Challenges like these were reflected in the reasoning of a third-grade class studying local aquatic systems. Students noticed that parts of the streambed in a nearby creek were occupied by abundant crayfish, whereas in other areas of the stream the crayfish were entirely absent (Lehrer & Schauble, 2017). Children also noticed that the composition of the streambed varied from place to place and proposed that perhaps the crayfish were deliberately choosing to live in those habitats that best hid them from predators.

To help students investigate their conjectures, the teacher installed a wading pool in her classroom to serve as a model of the stream. Students used masking tape to partition the pool into four equal quadrants. On the floor of each quadrant students replicated the kinds of substrates they had observed in the stream outdoors: one quadrant was covered with "mixed rocks," another with "white rocks," and a third with "mixed rocks with plants." The final quadrant was left empty. Ten crayfish were released into this environment, and over the next few weeks children observed the pool twice a day and recorded the number of crayfish found in each quadrant.

After the data were summed and displayed across all the observations, the students noted that the quadrant associated with the largest number of crayfish observations was the quadrant with plants. They readily concluded that the crayfish clearly preferred mixed rocks with plants and further proposed that this was because the plants provided protection from potential predators. At this point, the teacher challenged the students to consider an alternative explanation. Maybe, she proposed, crayfish did not actively prefer a particular kind of substrate. Perhaps the crayfish had simply wandered more frequently into that section of the model pond by chance. Students were reluctant to entertain this possibility; in their view, living organisms are intentional, and if they move from place to place, they do so deliberately. The teacher suggested that the class explore the behavior of some simple chance devices to see if the behavior of those devices could fairly represent the observed behavior of the crayfish.

At the teacher's suggestion, children first investigated outcomes from spinning a two-colored spinner divided into equal-area sections. After recording and discussing outcomes from multiple trials with the two-color spinner, they then transitioned to a four-section spinner intended to represent the four quadrants of the model pond. As children ran and recorded trials with both spinners, they struggled with many of the conceptions about chance that mathematics educators have previously noted (Horvath & Lehrer, 1998; Metz, 1996, 2010). Initially, they tended to either overestimate or underestimate what they could reasonably know in advance about the outcomes of a single spin. Some students asserted that they could control outcomes by varying the direction, starting point, or velocity of the spin, beliefs that they surrendered only after examining the data from trials in which half the class spun only to the right and half to the left or half spun "gently" while the other half spun "really fast." Some students began to suspect that these variants in "spin procedure" did not seem to substantively affect the pattern of outcomes, and these conversations helped anchor the important concept of trial (Horvath & Lehrer, op. cit.). Although no spin could be said to be identical to another (a little bit more or less force in the flick of the spinner, a change in the direction of rotation, etc.), students eventually concluded that spins could be treated as equivalent if the outcome of the spins could not be controlled.

As they reviewed the outcomes of these trials, a few youngsters vacillated between an overemphasis on determinism and an overemphasis on uncertainty (Biehler & Steinbring, 1991). That is, these children continued to insist that they could control outcomes by varying the spin, whereas most felt there was no basis at all for expecting any outcome to be more likely than any other (as one of the children concluded, "It could land anywhere, really"). However, as students combined their data from increasing numbers of trials (the process of growing a sample: Bakker, 2004; Horvath & Lehrer, op. cit.), nearly every student began to agree that, although it was not possible to predict the outcome of any particular spin, nonetheless, given many spins, some results were more likely than others. For instance, given 20 total spins, "5 green, 5 yellow, 5 red, 5 blue" was a result considered "more likely" than "20 green, 0 yellow, 0 red, 0 blue," even though the second result was considered "still possible."

Students next used uncolored spinners, divided into quarters, to create their own personal "mystery wheels" by coloring in the sections with either two, three, or four colors. Students spun their mystery wheels 20 times and then recorded and displayed the results in graphs. The class reviewed the outcomes and made predictions that linked graphs of the outcomes to the design of the spinner that produced them. From these investigations, students appeared to link the structure of a spinner to the pattern of outcomes over many trials. This conception of long-term process underlies the measure of chance as a probability (Saldanha & Thompson, 2002; Watson & Moritz, 2000).

After the spinner experiments were concluded, the teacher sought to connect children's emerging ideas about chance back to the original question about crayfish preference. She first heightened the distinction between intentional choice and chance by placing a different snack in each corner of the classroom and then presenting a spinner with quadrants labeled by small pictures of crackers, grapes, broccoli, and cookies, respectively. Each student spun the spinner and then went to stand in the corner associated with his or her outcome. The results were recorded and graphed; about an equal number of students ended up standing in each corner. Next, students were asked simply to go to the corner that represented the snack they personally preferred. As the teacher expected, most students chose cookies. These results were also recorded and displayed. The two displays initiated a discussion about the distinction between preference and chance, a distinction that in this case was dramatic.

Finally, children reconsidered their original question—that is, did the crayfish occupy particular locations in the model pond simply by chance, or did they actively prefer some substrates over others? The pattern of the data of crayfish location was more similar to the pattern that reflected cookie preference than to the pattern that reflected indifference to snacks, so students concluded that preference, not chance, better explained the observed distribution. A critical test of preference and function was proposed by one student, who suggested submerging green flowerpots turned upside down, with an entry cut into each, into one quadrant of the model pool. If crayfish preference was simply a matter of avoiding predators, then that quadrant would be where most crayfish would locate. If preference was a matter of camouflage, then the brown substrate areas would be most attractive. As in the previous example, distribution of organisms in space was attributed primarily to cause, but in the third grade, these causes were more firmly juxtaposed with the possibilities for variability generated by chance. Teasing apart variability due to chance and variability due to other processes confronts all ecological investigators (Albert et al., 2010), and this investigation gave these children some first-hand experience of these distinct potential sources of variability in an ecosystem.

Melding Chance and Cause: Investigating Precision of Measure

Although acts of measure rely on imagined ideals, in the world measuring is always accompanied by error, so all measures are somewhat uncertain. The extent of uncertainty is measured by the variability of repeated measures, also called a measure's precision, so any single measure can be considered as a composition of true measure (e.g., the length of a crab's carapace, the percent area of a pond covered by duckweed) and error. Understanding the precision of measure aids interpretation of differences in values of a measured attribute in that these differences can be considered in light of the precision of measure—the distribution that represents a measure's variability. Hence, measure is itself an example of a data-generating process that produces variability.

It is helpful to situate investigations about precision of measure in contexts where students can participate as agents of measure so that sources of variability in measure are within their grasp. These tangible processes are often characterized by signal and noise (Konold & Lehrer, 2008).

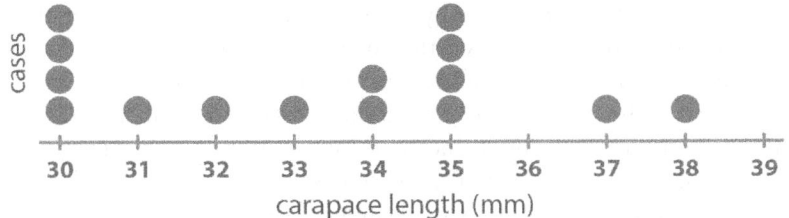

FIGURE 10.9 Frequency display showing that repeated measure of the same organism generates variability

For example, Figure 10.9 displays measures of the same crab's carapace. The measures were conducted by 15 students participating in a citizen science investigation of invasive species in intertidal zones in Maine (Wisittanawat, Dickes, & Lehrer, 2022). The variability of the measured values of carapace length surprised the students, who suggested multiple potential sources of variability. These sources included students' propensity to bias their reading of the ruler toward prominently marked landmarks (30 mm or 35 mm) and efforts by the crab to wiggle out of the grasp of its measurer, inducing random slips in measure.

Contexts of repeated measure not only make variability-generating processes tangible; they also create the need to measure characteristics of distribution to estimate true value and error. For instance, fourth-grade students investigated how the "spread" of their measures changed as each of them measured objects with tools that produced greater and lesser precision of measure (Petrosino, Lehrer, & Schauble, 2003). To begin, each student used a ruler to measure the length of a pencil and recorded the value. Subsequently, the class went outdoors, and each student measured the height of the school's flagpole with a commercially available plastic tool called an AltiTrak™, an altitude finder designed for measuring the heights of model rocket flights. Once the measures for the entire class had been plotted on frequency charts, students noted that the distributions of measures for both the pencil and the flagpole showed values clustered toward a center, with fewer cases toward the boundaries of the distribution—the classic normal shape. However, students found it striking that the pencil data were tightly clustered toward the center of the distribution, whereas the flagpole data were "more spread out." As they mulled over this difference, students proposed that the pencil was relatively easy to measure, and although it was possible to misread the ruler value that indicated where the pencil ended, those errors tended to be small and equally under- and over-shot the most frequently recorded value of 19 cm. The flagpole data were more variable, even though the shape of this distribution was also approximately normal. Students reasoned that the flagpole was more difficult to measure than the pencil because reading the

AltiTrak™, which was operated by aiming it at the top of the flagpole, was more difficult than laying down the ruler and reading the result. In both cases, qualities of students' data had readily interpretable correspondences in their own activity. They interpreted the center value (which was also the mode for both distributions) as the "true" measure of the object in question and thought of the other measures as representing components of error— error produced by shifts in individual students' care while measuring as well as by the tools that were used.

The teacher next asked students if it would be possible to describe the "amount of spread" in the data quantitatively and invited them to think of ways to do so. As they inspected the flagpole frequency graph on the board, a few students proposed that the range of the distribution could serve as such a measure. In response, the teacher asked how the range would change if someone had made a large error while reading the AltiTrak™. To illustrate, he added a hypothetical extreme value far to the right tail of the distribution, which was displayed as a frequency graph on the board. As he added this value, students saw that a mistake of this kind made by one student would greatly expand the range of the distribution. To most, it seemed unsatisfying that the spread of the distribution would change so dramatically even as the density of most of the measures remained about the same. Students next proposed that the range would work OK as a measure of spread if outliers were eliminated. This raised the problem of how to decide what counted as an outlier. Eventually the students settled on eliminating the top and bottom one-third of the data, calculating the range only on the middle two-thirds of the data.

Other students thought of spread a little differently, focusing not on the width of the entire distribution but, rather, on how far each individual value was from the distribution's center. Students proposed quantifying this approach by finding the distance of each score in the distribution from the middle value (the median/mode, which students interpreted as the "true" measure) and then summing all those distances. Although this solution aligned with students' views of "amount of deviation from center" as an indicator of precision of measure, students saw that the measure fluctuated as the distribution included fewer or more cases, even though intuitively the "error" in measure did not seem to change notably as the number of cases increased. Eventually students settled on creating a distribution of deviation scores from the median and then identifying the *median* of those distance scores. They dubbed this value the "spread number" of the distribution and interpreted it within this context as a measure of precision (a smaller spread number indicates better precision).

Children went on to investigate whether their spread number aligned with their intuitions that more and less precise tools for measuring the same height would yield different distributions of results. Students remeasured the height of the flagpole with a handmade cardboard tool and compared the spread number of that distribution of measures with the spread number of the existing distribution of measures made with the commercially produced AltiTrak™. Students' intuitions were confirmed when the AltiTrak™ tool resulted in a denser distribution with a smaller spread number.

Having established these ways of thinking about variability in distributions of measures, the fourth graders next proceeded to investigate whether and how design features of model rockets, such as the shape of the nose cone and type of engine, would affect the rockets' launch heights over a number of trials. Initially students confidently predicted that rockets with pointed nose cones would fly higher than those with rounded nose cones because the pointed shape would help the rockets "cut through the air." To investigate this conjecture, the 22 students conducted 37 trials with the rounded cone and 25 more with the pointed cone. At each trial, an observer measured and recorded the height of the launch at its apex, using the AltiTrak™. Students tended to think of variations in heights as occurring due to chance fluctuations in wind shear, measurement error, and so on.

When all the trials were completed and recorded, students displayed the resulting measures and noted that there was substantial overlap between the two distributions of measures. The median heights for the two distributions did not differ by very much, and students were unsure how to decide whether the distributions could be considered "really" different or, alternatively, whether the differences in the two distributions were due merely to measurement error and chance.

To interpret their findings, students combined their earlier ideas about partitioning the distribution with their later strategy for finding a spread number that described median deviation of scores from the center value. Students first divided the distribution for rounded nose cones into three "superbins." The middle superbin encompassed all the cases between the median and the spread number of that set of values, both below and above the median. The remaining two superbins comprised measures at the low end and the high end, respectively, of the distribution. Students calculated that 51% of the 37 rounded nose cone measurements fell within the middle superbin, 21% fell within the lower bin, and the remaining 27% appeared in the upper bin.

The teacher then suggested that students plot the pointed nose cone data on the same display, using different colored Xs to highlight the cases that represented the pointed nose cone launches. This strategy framed the original data for the rounded nose cone as a reference distribution. When they plotted the pointed nose cone data in the rounded nose cone reference superbins, students were surprised to see that 86% of the heights of the pointed nose cone rockets fell into the lower superbin and 14% into the middle superbin. None of the points fell into the high superbin. Students reluctantly discarded their original conjecture—that pointed nose cones would help rockets cut through the air and thus fly higher.

The context of measurement provided support for students' interpretations of statistical concepts related to distribution (Konold & Pollatsek, 2002). Measures of center corresponded to their sense of true scores, and measures of spread were interpreted as being due to the tools and techniques employed by the measurers. The overall shape of the distributions also made sense to students, given the nature of error in these contexts: people might easily measure a little (or in the case of the flagpole, a lot) over or under the true value. The students' spadework on qualities of measurement distributions set the stage for considering systematic effects (e.g., round vs. pointed rocket nose cones) in light of random effects due to error. Although even middle school students and adults are often reluctant to accept data that disconfirm their prior beliefs (e.g., Kuhn et al., 1988), these analyses were readily accepted by the class as disconfirming their prior theories about rocket design.

Modeling Measurements as Signal and Noise

Repeated measures provide entrée to a tangible variability-generating process and, in addition, are contexts in which statistics of center and variation are intelligible. For a repeated measure, statistics of center estimate the true value of a measure, and statistics of variability estimate the precision of the measure (Konold & Pollatsek, op. cit.). An additional instructional elaboration that was next pursued was to engage students in the invention and revision of models of the repeated-measure process. For example, Figure 10.10 depicts a facsimile of a model constructed by sixth-grade students to represent the process of measuring the arm span of their teacher (Lehrer, Schauble, & Wisittanawat, 2020). Each student used a 15-cm ruler to measure the teacher's outstretched arms, and all the measures were recorded. The model, based on these measures, was constructed with TinkerPlots™, a student-centered software package that provides an array of possibilities for representing variability-generating processes (Konold & Miller, 2006).

In the model depicted in Figure 10.10, the first device, the spinner, uses the median of the empirical sample (that is, of the collection of measurements made by the 30 students). In its simplest version, the model represents the measurement process as generating the median measured value in the sample. This model has the virtue of producing a value within a cluster of the measured values observed, but it mismatches the sample in that running the model would not produce any variability in the measures. Yet in fact, as students iterated the ruler, they noticed a tendency for slips of two kinds. Sometimes they inadvertently left gaps and underestimated the arm-span measure, while

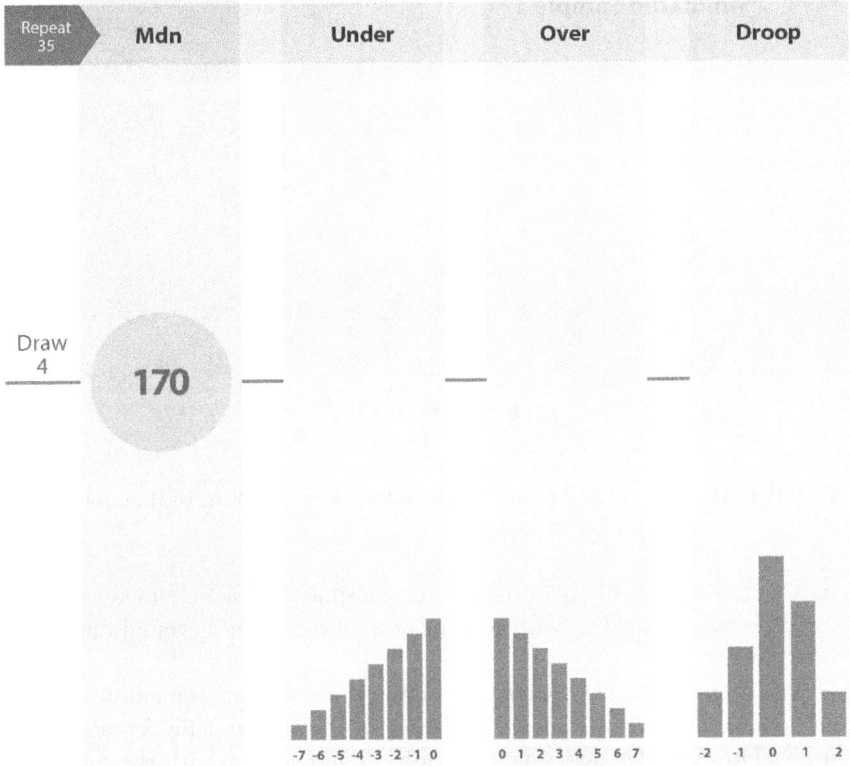

FIGURE 10.10 Modeling a repeated-measure process

at other times they accidentally overlapped the beginning of the next iteration with the endpoint of the previous iteration, thus overestimating the measure. After considering these problems, students decided that smaller magnitudes of error were more likely than larger magnitudes, and the "value bar" devices labeled *under* and *over* correspond to this analysis of the process. Each time the model is run, there is some likelihood of under- or overestimating the signal value (the median). In the model, larger magnitudes of errors are less likely, and smaller or no error due to iteration is more likely. The last value bar device represents variability caused by involuntary movement by the teacher due to fatigue as students repeatedly measured her arm span. The students called this source of error *droop* and experimented with different kinds of droop to develop an estimate of its contribution. During each trial, the model generates a simulated measure by adding randomly generated values of *under*, *over*, and *droop* to the median value. A simulated sample, in which the model was set at 35 repetitions to represent each member of the class (students and visitors), is

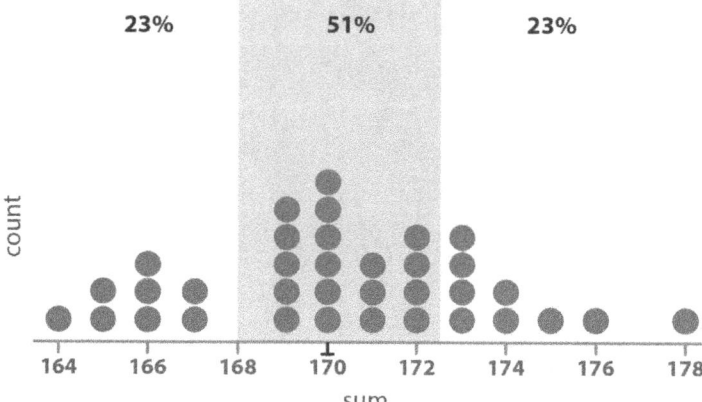

FIGURE 10.11 A simulated sample obtained by 35 repetitions of the model of the measurement process

depicted in Figure 10.11. In the figure, the simulated sample median is indicated by the inverted T, with percent of measured values around the median indicated.

The ability to decide whether a model is a good representation of a process is advanced when students recognize that as the model is repeatedly run, it generates samples that vary randomly in accordance with the underlying process. For example, Figure 10.12 represents another simulated sample of the same process depicted in Figure 10.11. Model fit offers students a route to

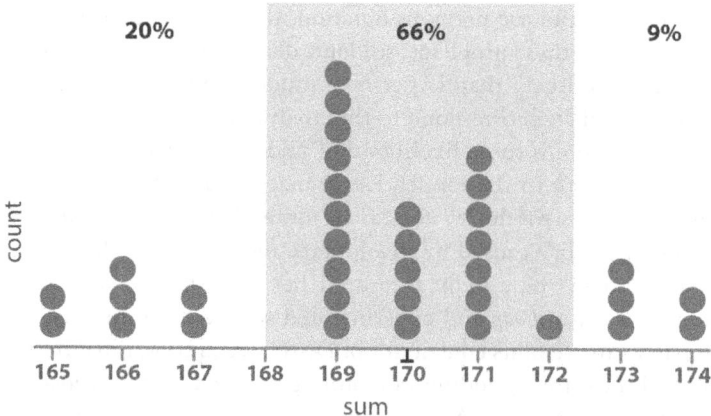

FIGURE 10.12 A different simulated sample of the same process

take sample-to-sample variation into account. For example, students consider whether simulated sample medians tend to cluster about the empirical sample median and whether the middle 50% of simulated values tends to have a range (the interquartile range) similar to the empirical sample's interquartile range over many runs of the model (each run produces a simulated sample). These criteria, in turn, help students contemplate whether the process depicted by the model is plausible or whether lack of fit suggests that some important component of the process may have been omitted.

The generation of many simulated samples creates grounds for model fit that are secured to images of the variability-generating process, but also creates a productive tension. Some simulated samples inevitably include simulated measures that do not have counterparts in the empirical sample. Observing these values raises a quandary: is the legitimacy of a model threatened if it generates values that are not observed in the empirical sample? If students are convinced that the model is a good representation of the repeated-measurement process, they tend to argue that even though the simulated value is not in the empirical sample, if the process were repeated (e.g., if another class measured the arm span or the crab's carapace), simulated values like these would nonetheless be "possible." Conceiving of possible values in this way creates a pathway for re-envisioning an empirical sample as only one of an infinite number of samples that could be generated by a particular process. This nested image of sample is important for statistical inference (Saldanha & Thompson, 2002), and repeated measure provides a tangible entrée to this form of reasoning (Lehrer, 2017).

Expanding the reach of modeling chance

As noted earlier, students participating in a citizen science investigation of invasive species in intertidal zones were first introduced to sample variability through repeated measure of a crab's carapace. As the scope of their inquiry expanded, students surveyed invasive crab abundance in 16 one-meter-squared quadrants sampled along a 60-meter transect (Wisittanawat et al., 2022). Unexpectedly, 88% of the 26 captive crabs were male, which did not seem consistent with students' prior belief that "in the ocean" there were equal numbers of males and females. Their teacher wondered if the observed ratio might vary "just by chance," as carapace length had for the repeated measures, so that perhaps the ratio observed in the field might not be as extraordinary as it seemed. If the survey process were repeated daily, the observed ratios might vary just by chance, even if the expected percentage of males in any sample was 50.

To investigate this possibility, students conducted sampling simulations with different sample sizes and ultimately decided upon a simulation with a sample size of 26 (to mimic the sample size in their field investigation), an equiprobable model of ratio (50% male), and a large number of imagined samples (e.g., in one group, 365 sampling events to correspond to an imagined extension of sampling to every day of the year). Students found that their empirical sample's sex ratio was improbable in light of the distribution of simulated samples, each of which was represented by a measure of percent male. Accordingly, students then turned their attention to empirical samples collected by fellow citizen science investigators at other intertidal locales. They noticed that the collection of sample statistics (percent male) in those locations was more consistent with their own sample and revised their model to better account for the possibility of random variability in sex ratios by favoring a higher percentage (71%) of males in the wild. This higher percentage of male, now taken as a robust finding, instigated a search for an ecological mechanism that would account for it. As students pursued an explanation, a few noticed that sex ratios appeared to change with the seasons, with higher male percentages more dominant in the fall (when they had sampled). This raised the possibility that sex differences in molting might produce bias in a sample (e.g., males had molted and hence were less vulnerable to predation in the portion of the intertidal sampled, and so were more abundant in the students' sample). For this reason, the findings might not reflect a change from an equiprobable sex ratio in the population ("in the ocean"). Alternatively, perhaps there were seasonal fluctuations in sex ratios in the wild and these fluctuations were not merely a matter of when and where the sample was collected. Conversations with partner scientists revealed that marine ecologists have not yet resolved these alternatives. Later in the school year, the same students conducted other sampling investigations to help them consider the relation between the number of samples collected and the probability of not observing a species in a vernal pool. These relations were considered in light of an estimate of the species' likely occurrence in a sample based on a larger collection of citizen science samples of vernal pools. Over the course of the year, students' thinking about random sample variability, initiated by unexpected differences in the measure of the same crab's carapace, was extended to thinking about sample-to-sample variability in service of inquiry about local ecosystems.

Modeling Broadens the Scope of Measures

At the beginning of this chapter, we asserted that measuring requires deploying and coordinating acts of the imagination. The episodes described here illustrate how the sciences, with their indissoluble connection to the empirical world, require new and broader dimensions of measure that are accompanied by the demand for new acts of the imagination. Measuring and interpreting observations of the natural world require new mathematical systems and forms of display. Cartesian coordinates and distributions are particularly productive for making sense of scientific phenomena, and their foundations can be brought within the grasp of young students. Along with these new mathematical ideas and displays come new forms of measure. These include ratio measures, such as rates to describe growth and density, to describe material kind, and statistics, such as center and spread, to describe qualities of distributions. Chance plays a prominent role in many models of natural phenomena, so measures of chance and the related concept of precision of measure also increasingly come into play as models and measures are evaluated.

Children may initially find these ideas to be disconcerting. For instance, school instruction does not always prepare students to expect that repeating a procedure may produce varying outcomes. Chance is by its very nature disconcerting; humans are inclined to seek definitive causes, and it is surprising that chance outcomes can nonetheless collectively have a structure. Finally, an appreciation for the very process of modeling must be cultivated. It is not initially evident why a model, which deliberately omits information about the world, may be a more useful representation than a veridical copy.

Nonetheless, these ideas, which may initially seem counterintuitive, are critical conceptual tools for participating in the generation and testing of scientific knowledge. Without them, students are constrained to distortions of scientific inquiry that involve following procedures determined by someone else to produce findings that are predetermined. As they achieve some purchase on models and measures, including an understanding of how models and measures are interrelated, students develop enhanced potential to pursue their own questions about the world and to interpret what the world has to say.

References

Albert, C. H., Yoccoz, N. G., Edwards, T. C., Graham, C. H., Zimmerman, N. E., & Thuiller, W. (2010). Sampling in ecology and evolution—bridging the gap between theory and practice. *Ecography*, *33*, 1028–1037.

Bakker, A. (2004). Reasoning about shape as a pattern in variability. *Statistics Education Research Journal*, *3*(2), 64–83.

Biehler, R., &, H. (1991). Statistics by discovery, stem-and-leaf, box plots: Basic conceptions, pedagogical rationale, and experiences from a teaching experiment. *Der Mathematikunterricht*, *37*(6), 5–32.

Cannaday, M. A., Vincent-Ruz, P., Chung, J. M., & Schunn, C. D. (2019). Scientific sensemaking supports science content learning across disciplines and instructional contexts. *Contemporary Educational Psychology*, *59*, 1–15.

Chang, H. (2004). *Inventing temperature: Measurement and scientific progress*. Oxford: Oxford University Press.

Cobb, P., & McClain, K. (2004). Principles of instructional design for supporting the development of students' statistical reasoning. In D. Ben-Zvi & J. Garfield (Eds.), *The challenge of developing statistical literacy, reasoning, and thinking* (pp. 375–396). Dordrecht: Kluwer Academic.

Horvath, J., & Lehrer, R. (1998). A model-based perspective on the development of children's understanding of chance and uncertainty. In S. P. LaJoie (Ed.), *Reflections on statistics: Agendas for learning, teaching, and assessment in K–12* (pp. 121–148). Mahwah, NJ: Lawrence Erlbaum Associates.

Kanari, Z., & Millar, R. (2004). Reasoning from data: How students collect and interpret data in science investigations. *Journal of Research in Science Teaching, 41*(7), 748–769.

Klahr, D., Fay, A. L., & Dunbar, K. (1993). Heuristics for scientific experimentation: A developmental study. *Cognitive Psychology, 24*(1), 111–146.

Knapp, N., & Lehrer, R. (2005, June). Changes in children's conceptions of spatial measure: Coordinating talk and inscription. In M. Wiser (Organizer), *Understanding, building, and using symbolic representations of space and time*. Paper presented at the 35th Annual Meeting of the Jean Piaget Society, Vancouver, Canada.

Konold, C., & Lehrer, R. (2008). Technology and mathematics education: An essay in honor of Jim Kaput. In L. D. English (Ed.), *Handbook of international research in mathematics education* (2nd ed.), pp. 49–72. Philadelphia: Taylor & Francis.

Konold, C., & Miller, C. D. (2006). *TinkerPlots: Dynamic data exploration* (computer software). Emeryville, CA: Key Curriculum Press.

Konold, C., & Pollatsek, A. (2002). Data analysis as the search for signals in noisy process. *Journal for Research in Mathematics Education, 33*(4), 259–289.

Kuhn, D., Amsel, E., & O'Loughlin, M. (1988). *The development of scientific thinking skills*. San Diego, CA: Academic Press.

Lehrer, R. (2000, April). Designing for development. In E. Forman (Chair), *Strengths and challenges of developmental approaches to research in education*. Paper presented at the Annual Meeting of the American Educational Research Association, Seattle, WA.

Lehrer, R. (2017). Modeling signal-noise processes supports student construction of a hierarchical image of sample. *Statistics Education Research Journal, 16*(2), 64–85.

Lehrer, R., & Schauble, L. (2000). Inventing data structures for representational purposes: Elementary grade students' classification models. *Mathematical Thinking and Learning, 2*(1&2), 51–74.

Lehrer, R., & Schauble, L. (2004). Modeling natural variation through distribution. *American Educational Research Journal, 41*(3), 635–679.

Lehrer, R., & Schauble, L. (2012). Seeding evolutionary thinking by engaging children in modeling its foundations. *Science Education, 96*(40), 701–724.

Lehrer, R., & Schauble, L. (2017). Children's conceptions of sampling in local ecosystem investigations. *Science Education, 101*(6), 968–984.

Lehrer, R., Schauble, L., Carpenter, S., & Penner, D. (2000). Symbolic communication in mathematics and science: Co-constituting inscription and thought. In E. Amsel & J. P. Byrnes (Eds.), *Language, literacy, and cognitive development: The development and consequences of symbolic communication* (pp. 167–192). Mahwah, NJ: Lawrence Erlbaum Associates.

Lehrer, R., Schauble, L., & Lucas, D. (2008). Supporting development of the epistemology of inquiry. *Cognitive Development, 23*(4), 512–529.

Lehrer, R., Schauble, L., & Wisittanawat, P. (2020, August). Getting a grip on variability. In Mathematical Biology Education, special issue of the *Bulletin of Mathematical Biology, 82*, 106.

Lehrer, R., Strom, D., & Confrey, J. (2002). Grounding metaphors and inscriptional resonance: Children's emerging understanding of mathematical similarity. *Cognition and Instruction, 20*(3), 359–398.

Levin, I., Israeli, E., & Darom, E. (1978). The development of time concepts in young children: The relations between duration and succession. *Child Development, 49*(3), 755–764.

Manz, E., Lehrer, R., & Schauble, L. (2020). Rethinking the classroom science investigation. *Journal for Research in Science Teaching, 57*(7), 1148–1174.

McClain, K., & Cobb, P. (2001). Supporting students' ability to reason about data. *Educational Studies in Mathematics, 45*, 103–129.

Metz, K. E. (1996). Emergent understanding and attribution of randomness: Comparative analysis of the reasoning of primary grade children and undergraduates. *Cognition and Instruction, 16*(3), 285–365.

Metz, K. E. (2010). Children's understanding of scientific inquiry: Their conceptualization of uncertainty in investigations of their own design. *Cognition and Instruction, 22*(2), 219–220.

Nersessian, N. J. (2008). *Creating scientific concepts.* Cambridge, MA: The MIT Press.

Petrosino, A., Lehrer, R., & Schauble, L. (2003). Structuring error and experimental variation as distribution in the fourth grade. *Mathematical Thinking and Learning, 5*(2&3), 131–156.

Pickering, A. (1995). *The mangle of practice: Time, agency, and science.* Chicago, IL: University of Chicago Press.

Saldanha, L. A., & Thompson, P. W. (2002). Conceptions of sample and their relationship to statistical inference. *Educational Studies in Mathematics, 51*(3), 257–270.

Schauble, L. (1990). Belief revision in children: The role of prior knowledge and strategies for generating evidence. *Journal of Experimental Child Psychology, 49*(1), 31–57.

Schauble, L., Glaser, R., Duschl, R., Schulze, S. & John, J. (1995). Students' understanding of the objectives and procedures of experimentation in the science classroom. *Journal of the Learning Sciences, 4*(2), 11–166.

Smith, J. P. III, Males, L., Dietiker, L. C., Lee, K., & Mosier, A. (2013). Curricular treatments of length measurement in the United States: Do they address known learning challenges? *Cognition and Instruction, 31*(4), 388–433.

Thomas, M., Clarke, D., McDonough, A., & Clarkson, P. (2016). *Understanding time: A research-based framework.* Paper presented at the 39th Annual Meeting of the Mathematics Education Research Group of Australasia (MERGA), Adelaide, South Australia.

Thompson, T. D., & Preston, R. V. (2004). Measurement in the middle grades: Insights from NAEP and TIMS. *Mathematics Teaching in the Middle School, 8*(9), 514–519.

Watson, J., & Moritz, J. B. (2000). The longitudinal development of understanding of average. *Mathematical Thinking and Learning, 2*(1–2), 11–50.

Wisittanawat, P., Dickes, A., & Lehrer, R. (2022). Modeling chance processes in a classroom's ecological investigation. In C. Chinn, E. Tan, C. Chan, & Y. Kali (Eds.), *Proceedings of the 16th International Conference of the Learning Sciences* (ICLS 2022) (pp. 719–772). International Society of the Learning Sciences.

INDEX

Italicized and **bold** pages refer to figures and tables respectively, and page numbers followed by "n" refer to notes

addition 54, 127

angle measure 6; by accretion and generalization 105; as attributes of figures and structures 108; benchmark levels of conceptual development 108–122; degree measure 112; dynamic and figural perspectives 104–108; example of own bodily experiences 106; exterior *113*, 113–114; generating measure theorems 116–120, **118**, 156–158; interior *113*, 113–114; Logo turtle example 105–107, 154; path perspective 154–155; path taken by Lego™ figure 154–155; representations of angle-as-figure and angle-as-turn 108, **109**, *110–111*, 110–112, **117**, 154; tasks for assessing students' understanding of 155–156; tools for measuring 114–116, *115*, 119; turn *112–113*, 112–116, *115*, *120*; understandings of figures and structures using 120–122, *121*

area measure 6, 67–68, *98*, 101, 149–154, *150–151*, *163*; Cavalieri's principle of equivalent areas 80, *80*, 82; comparing perimeter and *75*, 75–76; comparison of magnitude 68–72; conditions of 150; dissection

strategies 152; distinction between length and 152–154; dynamic generation of 76–83, *77–80*; of handprint 72–73, *74*; of non-closed forms 74, *74*; of parallelogram 82, *82*; of polygon 113, 119, *119*, 151, *151*; properties of units 72–76, **73**; of rectangles 69–70, 76, 81; reinvention of formula 83–86, *85*; scalar multiples 136; surface area 92; of triangle 84, *84*; unit structuring 151–152; variations of identical-part dissections 70, *71*, 72

assessment system 7, 141; Berkeley Assessment System Software (BASS) 29–30, 36; constructs and construct-centered design 11, 19–20, 22–24, 26–30, **27**, *31–32*, 36–38, 52–55, 57–59, 61, 68, 71–72, 80, 82–83, 85, 92–93, 98, 106–107, 112, 114, 116, **117**, 119–120, 124–128, 130–131, **132**, 133–135, 138, 141–142, 150–151, 154, 158–159, 161–162–169, 177–184, 186–190, 200–204, 207–209, 215–217, 222, 232; formative assessments 142, 164–173; in-situ construct-centered teacher observation of student learning 164–173, multi-tiered 141; summative

assessments 141, 161–164, *162–163*; teacher observation tools (TOTs) 29–30, *31*, 36, 142, 201–202
attributes, measure of 46–47

benchmark levels of thinking 43–64; angle 108–122; attributes 46–47; comparison of perceived magnitudes 44–48, **47**; concept of baseline 46; iteration 52–55; mixtures of units 50; practical activity 49; properties of units 48–52, **51**; pumpkin heights and girths 45; relationships among units and measures 61–64, **63**; standard ruler 60–61; symbolic structure 49
boundary filling 5

Cartesian coordinates 208–209
Cavalieri's principle: to establish equivalent areas 80, *80,* 82, 86 and volumes 160–161
Children's Measurement Project 8, 10–11
circular protractors 114–116, *115*, 156
classroom-based assessment 29
classroom-based design studies 33
classroom instructional activities 7
closure property 75
Cognitively Guided Instruction 9, 18
composite units 54
coordination classes 41
copying metaphor for unit iteration 55
crab's carapace, measures of 228–232, *229*, 235
cubes 92–93, *93*
curriculum development 24–26

design research 18–19, and design studies 19, 33, 150
dialogic space 167–173, *168, 172*
display of data 32
distance measure 4–6, 50
distributional thinking/distributional reasoning 209, 220–225, *222–223*

earth's circumference 1–2, *2*
educational design, elements of: assessment system 26–29, **27**; digital tools 29–33; generation and refinement of learning constructs 22–24; professional development collaboration 20–22, 21; supporting curriculum units 24–26

education innovation and research: designed context 18–19; educational design 19–33; research sites and participants 17–18
eight-splits of a unit length 129–131
embodied metaphors of measure: as bodily rotation to make/measure angles 106–107, 111, *111*; as copies/translations of rigid stick/body-part 11; as dynamic movement along path 5, 44, 54, 57–58, 72, **81**, 105, 107, *112–113*, 113–114, 116, 120, 126–127, 154–157, 165, 185, 188, 190, 196; as splitting to fracture units 129–138, **130**, **132**, **136**, *133–135*, *137*; as sweeping area through length to create volumes 89–90, 97, 99, **100**, 101, 169, 172–173, 190, 197; as sweeping one length through another length to create areas 34, 77–80, *78–79*, 82, *83*, 136, *137*, 150, 153, 163, 168
epistemic culture 175
equality and inequality 9
equilateral triangle as path 113, *113*, 157, 171
equipartitioning 55; and fractured units 148
Eratosthenes of Cyrene 1–2
exploratory learning studies 33–34

formative assessments 142, 164–173, *167*; conversation 167–173, *168, 172*; observations of student responses 165; *in situ* observations 165–166, *166*; tasks 34, 36; *see also* assessment system
formative assessment tasks 34, 36
four-splits of a unit length 129–131
fractions 9, 12, 57; multiplication involving 135–138; as operators 132–138, **136**

half-unit iteration 57, 128–129
humble theories 19

in situ observations 36, 165–166, *166*
intuitive and experiential knowledge 10
iteration 52–55, 125–126, 147–148

K-5 mathematics education 124

learning constructs, generation and refinement of 22–24; benchmarks 23

learning ecology 43
length measure 5–6, 12, *31–32*, 34, 42; augmenting 54; benchmark levels of 44–64; children's understanding of 43–44; direct comparison of different attributes 144–145; early-developing conceptions 143–149; equipartitioning fractured units 148; instructional steps for 164; magnitude 41–42, 50, 52, 55, 125–126, 128, *164*, 165; measurement arithmetic 148–149; motion of one length through another 81; as a point along a path 126; short and long lengths 50; of squashes (pumpkins) 164–165; of squeegee 80–82; sweeping lengths 77, *78*, 79, 82, *82–83*; tiled multiple units 145–147

magnitudes 44–48, **47**; of area 68–72, *71*, 72; distinction between measure and magnitude 126; of length 41–42, 50, 52, 55, 125–126, 128; of space 68; of a subdivided unit and its parent unit 124
mathematics curriculum 24
measurement 1; coordinated perspectives on 67–68; distinction between magnitudes and measures 126; importance in scientific inquiry 209; in science class 207; as signal and noise 232–237; tools 4
measurement models of arithmetic148–149; addition as joining lengths or continuing motion, 59; division as re-measure 136–137; multiplication as split-copy of measured quantities, 133–135, **136**
Measure Up project 8–10
modeling 208
multiplication 9; commutative property of 79, *79*; distributive property of 79, *79*
multiplicative comparisons of quantities 54–55, 59
multi-tiered assessment system 141

natural systems 208

one-half unit 57

paces 121
parallelograms, area of 82, *82*
part-unit iteration 57

pentagonal prism 159–160
polar coordinates 121
polygons 113, 119, *119*, 151, *151*
precision of measure 228–232, *229*
prisms 170; of cubic units 158–159; with partial structuring 159
prisms, volume of 96, *96–98*, 98–99; of non-right 99; oblique 99; right 99
professional development 176; in contexts of shared activity structures 182–188; of a shared professional vision 177–182
professional development collaboration 20–22, 21
professional vision of teaching and learning 203–204

qualities of measure 210
quantitative reasoning: additive comparisons of quantities 42, 58–61, 86, 124, 127, 133, 138, 184, 187; multiplicative comparisons of quantities 59–61, 86, 124, 127, 138, 184, 187; quantity 4

rational number 12, 124–125; compositions of 2- and 3-splits, 131, **132**; development of reasoning with 125–127; equipartitioning and splitting of unit: 2-splits 55–59, **60**, 127–129, *129*, **130**; 3-splits 59–61, 131, **132**; 4- and 8-splits 129–131; as measured quantities 127–131; recursive partitioning 131; as scalar multipliers of measured quantities of length 135–138
reciprocal relations between A measured in B and B measured in A 59, *63*
rectangular area 6
rectangular area and volume 6
recursive partitioning 131
reproduction in measure 46
rhombus 171
rigid stick metaphor 11

scalar multiple 136–137
science education 208; crayfish observations 226–228; describing change by determining differences in quantity 210–212, *211*; differences in rates 212–216; distributional thinking/distributional reasoning 209, 220–225, *222–223*; measures of distribution 216–220, *217–218*; plant

growth 212–216, *213*; population changes 216–220; precision of measure 228–232, *229*; understanding of cause and chance 225–228; variability-generating process 222–223, 229, 232, 235

shared learning constructs and tools 182

spatial measure 5–6, 41; congruence of spaces covered by dissection *69*

spatial reasoning 10

split-unit measures: 2-split 55–59, **60**, 127–129, *129*, **130**; 3-split 59–61, **62**, 131, **132**; 4-split 129–131; 8-split 129–131; composition of 2- and 3-split, 131, **132**

square unit 74

student learning research 33, 161–173; ability/performance levels in mathematics 35; exploratory learning studies 33–34; formative assessment tasks 34, 36; measurement of length and angle 35; phases 140–173; reflections and prospects 173; in situ observations 36; yearly interview data 34–36, **35**, 142–143, **143**

subtraction 127

summative assessments 141, 161–164, *162–163*

swept space 34

symbolization 126–127, 138; of half-unit 128

symmetries 121

tape measures 45–46, 49, 53–55, *54*, 57, 128–130, 147–148

teacher judgment 142

teacher learning and practice 36–38; benchmark levels and exemplary performances 38; construct-centered judgments 201–203; learning labs 198–199; mathematical investigations 199–200; noticings 189–197, **193**; observations of student learning 175; (*see also* assessment system); professional development 37, 197–201; professional vision 203–205; video episodes 189–191; views of communal critique 200–201

teacher observation tools (TOTs) 29–30, *31*, 142, 201–202; iPad system 36

teacher-researcher partnership 176

teachers' construct-centered judgments 201–202

theory of measure 2–7, 41; as an interchange between imagination

and pragmatic activity 2; contexts and situations of measure 4; as knowledge-generating activity 42; measure of length, conceptual system for 3; principle of measure 49; properties of units 3; relational thinking about measures and magnitudes 42; relations between iteration and measurement scale *3*; spatial domains 42; travel/motion metaphor 5–6, 9, 49

three-split units 59–61, 131, **132**

TinkerPlots™ 219, 232

transversal of parallel lines 121

travel/motion metaphor 5–6, 9, 49

triangle, area of 84, *84*

turn angles *112–113*, 112–116, *115*; of regular hexagon 157; *see also* angle measure

two-split units 55–59, **60**, 127–129, *129*, **130**

units: iteration and copying 52–55, **56**, 125–126, 147–148; properties of 48–52, **51**; relationships among measures and 61–64, *63*; *see also* split-unit measures

unit-scale construction 52–55, *54*

variability-generating process 222–223, 229, 232, 235

volume measure 12, 158–161, 168; Cavalieri's principle 160–161; as composed of layers 93–96, *94–95*; connections between area and 101; dynamic generation of 97–101, **100**; fractional measures *137*; in fractional units 96, *96–97*, **97**; interrelated dimensions 97; as lattice of cubes 92–93, *93*, 95, *95*; movement of base 99; multiplicative coordination of dimension 99; of pentagonal prism 159–160; of prisms 96, *96–98*, 98, 158–159, 170; as a product 98; properties of units 91–93, **94**; scalar multiples 136; as space inside 90–91, **91**; structuring and dynamic approaches to 88–90

Wisconsin Fast Plant™ 212–216, *213*; day of growth 214, *214*; form of growth curve 213, *213*; root and shoot growth 215, *215*